To my good friend Chris

Since you always listen to me)

we are human counselling
PUBLISHING

wearehumancounselling.com

Copyright © We Are Human Counselling Limited. All rights reserved. No part of this book shall be reproduced, stored in a retrieval system, or transmitted by any means—electronic, mechanical, photocopying, recording, or otherwise—without written permission from the publisher, except for the inclusion of brief quotations in a review.

Every effort has been made to make this book as complete and as accurate as possible, but no warranty of fitness is implied. The information is provided on an as-is basis. The authors and the publisher shall have neither liability nor responsibility to any person or entity with respect to any loss or damages arising from the information contained in this book.

Dedication

I would firstly like to thank all the participants that took part in this research. Without them, this research would have been impossible. I would also like to thank many of the academic team at Roehampton, who have been my support throughout this journey. Most of all, I would like to thank my academic supervisors, Dr Gella Richards, for her constant support and belief in my abilities—thank you. I would like to further show gratitude to the individuals who encouraged me to write this book, thank you.

I would also like to extend my appreciation to the women who helped me through my counselling psychology doctorate, Ms Jo Cruywagen, Dr Jean O'Callaghan, Professor Marcia Worrell, and finally Dr Diane Bray, who has always believed in me. Lastly, I would like to thank my lovely parents, especially my beautiful mother. You are my inspiration.

Dr Farah Nadeem

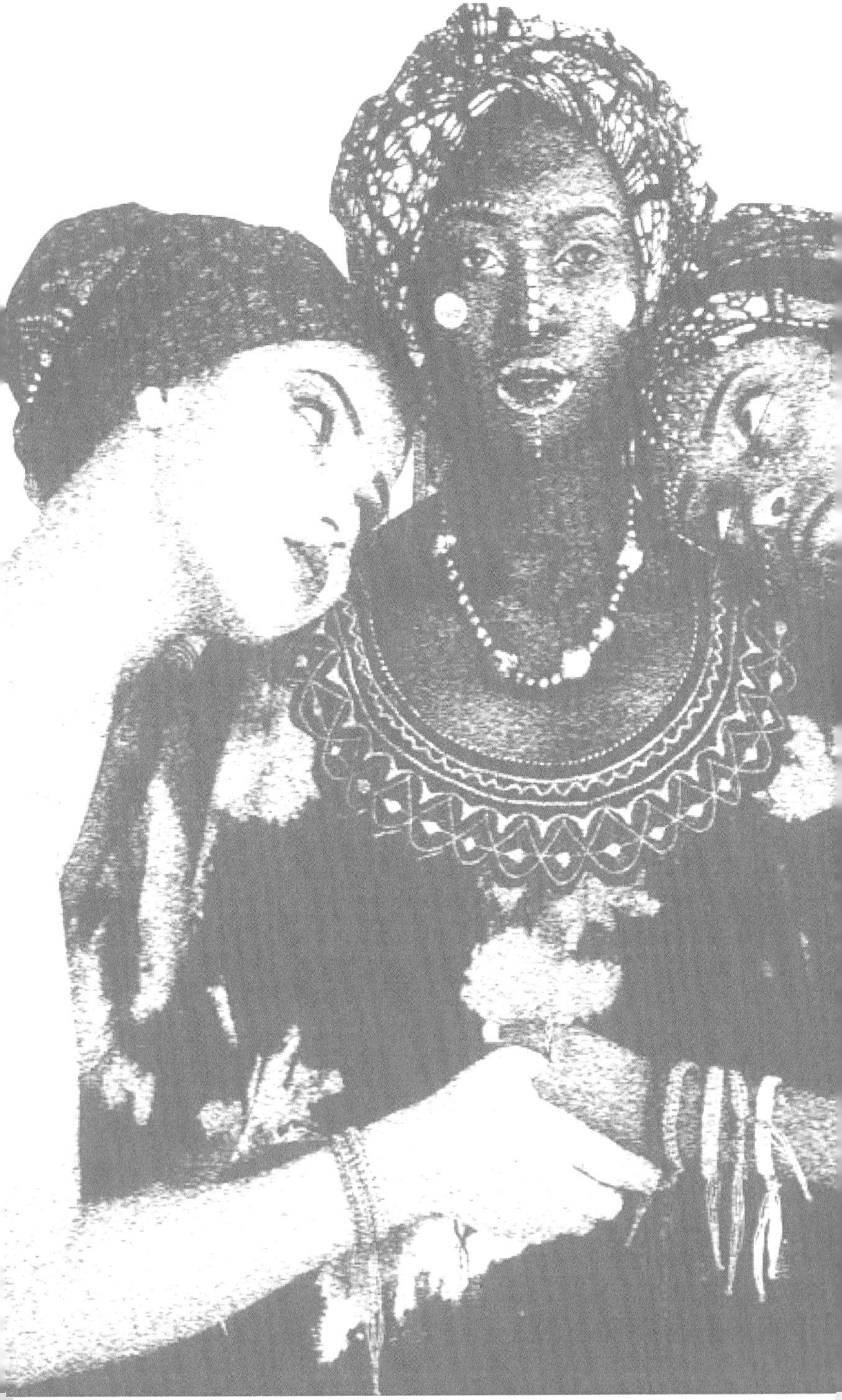

Contents

i	Foreword
v	Acknowledgments
18	Introduction
29	Global impact of FGM
35	Perspectives on FGM
46	Working with Trauma in counselling
62	Pen portraits of participants
73	The psychological impact of working with FGM
96	Emotional impact of working with FGM
106	Cultural Dynamic in the work with FGM clients
127	Therapeutic implications regarding the work with FGM survivors
150	Reflections and learnings when working with FGM
161	Implications for clinical practice
166	Closing remarks and reflections
178	References

Foreword

Counselling psychologists experience many different clinical presentations within their therapeutic practice. Understanding these experiences is essential to developing awareness and learning from issues that can arise when working with clients. The present book will look at the psychological and emotional impact therapists may experience when working with survivors of Female Genital Mutilation (FGM). The current cohort used for this study was investigated between the years of 2015-2016.

As the researcher of this study, I feel it is important to provide a brief introduction regarding my passion for work with FGM clients. This began when I encountered a client in the course of my clinical training for my counselling psychology doctorate. In our work together, the client chose to focus on her experience of FGM. I felt overwhelmed by many of the issues the client chose to disclose. The journey I took with this client led to a discovery of the lack of research regarding therapeutic work with FGM survivors.

Furthermore, many researchers and supervisors knew little about the impact this work may have on the therapist. There was also an overall dearth of knowledge regarding therapeutic interventions. Henceforth, this book aims to take the reader through the journey of seven psychological therapists/psychologists/counsellors (titles used interchangeably) and their lived experiences and meaning-making processes when working with clients who are FGM survivors.

Introduction: Female Genital Mutilation "FGM is the deliberate mutilation of female genitalia. This is often the removal or cutting of the labia and clitoris. The World Health Organisation describes FGM as any procedure that injures the female genital organs for non-medical reasons. FGM has no health benefits and is in fact very harmful to health in many ways" (Daughters of Eve, 2017).

Aim: The proposed research will explore the experiences of therapists who have worked therapeutically with adult women who have undergone FGM in childhood.

Method: Interpretative phenomenological analysis (IPA; Smith, 2004) involving the analysis of seven semi-structured interviews with (a variety of) psychological therapists. Initially, individual transcripts were analysed separately, yielding a number of themes for each participant. A group analysis was then conducted across participants, yielding superordinate themes and sub-themes based on their psychological relatedness.

Results: Four superordinate themes (with associated sub-themes) were identified: 1) The psychological impact of working with FGM clients (vicarious traumatisation; parallels with clients' helplessness; identity vs. loss of self; positive growth). 2) The emotional impact of working with FGM (feelings of guilt, sadness and anger). 3) Cultural dynamics in work with FGM clients (cultural embeddedness of the act; language barriers; child abuse vs. context of the act). 4)

Therapeutic implications regarding work with FGM survivors (need for supervision; personal therapy; good vs. bad supervision). The findings demonstrated an approach to understanding the way therapists work with a form of trauma.

Conclusion: This research explicated the significance of understanding the consequences of working with FGM clients, particularly the application to counselling psychology regarding trauma work. The findings demonstrated potential contributions regarding ways that counselling psychologists/therapists/counsellors may work with this client group. In particular, understanding the cultural relativist view was essential to the work; having some experience with trauma work was also deemed very important for future therapists to reflect on. Furthermore, the implications of working with FGM clients led to the inference that self-care, in terms of the right clinical supervision, was essential in therapists' meaning-making.

Overall, this book provides clinicians with some insight into possible ways of working therapeutically with FGM survivors.

Acknowledgements

Professor Helen Cowie, Faculty of Health and Medical Sciences, University of Surrey

This is a very timely book on a topic that concerns millions of women and girls worldwide. Dr Nadeem discusses in depth the emotional impact on therapists who work with survivors who have suffered FGM in childhood and adulthood and offers perceptive insights based on her own research in this field. She gives moving accounts of the ways in which therapists can gain deeper understanding of the cultural contexts of the practice of FGM and can consider the best ways, including the sensitive use of language, in which to build trust with their clients Most importantly, this book indicates how effectively therapists can improve their practice through heightened awareness of the trauma aspect of working with FGM clients and their willingness to seek out appropriate supervision and training. This pioneering book is essential reading for all those who work with survivors of FGM.

Rt Hon Stephen Hammond MP
Member of Parliament (December 2018)

Female Genital Mutilation, or female circumcision, is serious abuse of females, generally those within Black Minority Ethnic (BME) communities, and has no place in British society. FGM has been illegal in the UK since 1985 and is punishable by up to 14 years in prison.

With an increase in migration to the UK of people who still practice this horrific custom, FGM has become more prevalent in the UK and it is estimated that approximately 137,000 women and girls have been victims of the practice.

This book By Dr Nadeem is composed to help all practitioners for the treatment of FGM. Dr Nadeem has investigated the way this work impacts therapist, and describes clinical implications to the work. This book is written to aid the future treatment of FGM.

It is very important that doctors, nurses, psychologist, therapist, midwives, social workers and teachers are trained to recognize the signs of FGM and report it to the police, and it is also vital to have experts in the field of FGM who can help educate communities and help women stand up to FGM. This book aims to help all that work in this field and educate all professionals.

I welcome Dr Nadeem's book as she is a leading advocate in this field aiming to provide better mental health care for FGM survivors, by helping practitioners understand issues to be mindful of when working with this client/patient group.

This is a fascinating read and I endorse this book for all to read when aiming to learn about themes that appear when working with trauma such as FGM.

* * * * * * * * * * * * * * * *

"Four women held me down. I felt every single cut. I was screaming so much I just blacked out"

(Hussain, 2013)

Dedicated to the sisters without voices...

Chapter I

An Introduction to FGM

Female genital mutilation (FGM) is a social phenomenon of which there is increasing awareness in the UK. FGM, which is considered a breach of human rights, is outlawed in a number of countries (Momoh, 2010). In the UK, FGM is considered child abuse and is an illegal practice (FGM Act, 2003).

Despite the past decade having seen increases in studies and recommendations for healthcare support related to the physical health consequences of FGM, little is known about its psychological impact or the management of treatment (Tobin & Jaggar, 2013). This book proposes to explore therapists' experiences of working with an adult who has undergone FGM in childhood. The reason for this book was to investigate the extent to which counselling and therapy services are equipped to deal with this issue in the counselling room. There are many conflicting cultural approaches to the practice and the ethics involved in the conduct of parties to such experiences (Bell, 2005). For example, each therapist brings their own cultural background and experience to their work. Furthermore, work with such clients involves family systems extending across national borders, with varying ethical approaches.

The book seeks to provide comprehensive, psychological knowledge of the subject, combined with a working understanding of its examination of underlying issues that emerge in the therapist/patient relationship. This book is outlined in 12 chapters, these chapters are taken from the thesis I completed for my practitioner doctorate to qualify as a psychologist. Chapter one is the introduction of the book which outlines a brief description of each chapter. Followed by the types of FGM and medical consequences.

This chapter also explores the reasons given for this practice, such as social and religious elements etc. It is important to note that is not an Islamic practice and is not part of the religion. Chapter 2 is a reflection of the prevalence and descriptions of female genital mutilation, exploring through history and statistics to help readers understand the current global impact of FGM. This chapter in particular, considers the UK epidemic of FGM. While, chapter 3 reflects on the perspectives on the practice of FGM. Due to the varied nature of these perspectives, reflections in the book will specifically be made from the feminist perspective, the cultural relativist view and the standpoint of medicalising FGM. Another viewpoint pertains to the psychological and social perspectives of women experiencing shame if the act is not performed (Hayes, 1975; Hosken, 1993).

Chapter four, then, provides the reader with knowledge of issues therapist/psychologist accounts when working with trauma in clinical work. Reflections on core challenges professionals' experience such as: burn out, vicarious trauma, and issues impacting the therapeutic relationship. Additionally, reviewing possible presentation issues among trauma clients.

Chapter five, then describes the participants involved in the study conducted, by giving pen portraits of each participants. This gives the reader a description of each therapist experiences of the work they have done with FGM patients/clients. Descriptions of secondary care, of group work and classic one to one therapist/patient relationship is described. Furthermore, describing their years of experience and the type of psychologist, therapist or counsellor they practice as and deliberating their therapeutic modality. This book begins to unfold the experiences clinicians have when working with FGM survivors.

This is organized by verbatim transcripts from the subject experience of each clinician, an interpretative phenomenological analysis (IPA) was completed to develop the themes that create chapters six to nine. Chapter six, focuses on the psychological impact of working with FGM survivors, exploring issues of: vicarious traumatization (Intrusiveness, burnout, trauma work), parallels with client's helplessness, identity vs loss of self and positive-growth.

Chapter seven, explores the emotional effect the work with FGM survivors has on psychological therapist. Themes revealed were guilt, sadness and anger/frustration. The sub themes of guilt indicated aspects of religious connotations, while sadness exposed associated guilt and feelings of bereavement. Chapter eight, describes the cultural dynamics within the treatment of FGM survivors. Uncovering the thoughts of the culture embeddedness of the act, language barriers one may experience, the emergence of safeguarding issues relating to child abuse vs the context of the act debate. Furthermore, issues of clinical risk are also explored in the narratives of clinicians.

Chapter nine, describes the therapeutic implications regarding the work with FGM survivors, deliberating aspects of the need for psychoeducation and important aspects of the work. Some differences were described in the theoretical and therapeutic modalities, issues with trust vs. privacy, elements important for future work with FGM clients and the importance of self-care and the need for supervision/personal therapy were discussed.

Chapter ten, reflects on learnings we can take from the experience of treating FGM patients and clients.

Data was collected through semi-structured interviews and analysed through IPA in order to address the following research question: 'How have therapists experienced working with clients who have suffered FGM in childhood, and how might such experience inform future guidelines and practices with this client group?', this chapter focuses on addressing this question and also analyses the findings.

Chapter eleven, then looks at the clinical implication for practitioners within the therapeutic field. Exposing the significance of understanding the consequences of working with FGM clients, particularly the application to counselling psychology regarding trauma work. The findings demonstrated potential contributions regarding ways that counselling psychologists / therapists / counsellors may work with this client group.

Understanding the cultural relativist view was essential to

the work; having some experience with trauma work was also deemed very important for future therapists to reflect on. The implications of working with FGM clients led to the inference that self-care, in terms of the right clinical supervision, was essential in therapists' meaning-making. Overall, this book provides clinicians with some insight into possible ways of working therapeutically with FGM survivors.

The last chapter, chapter 12, is closing remarks of the work, I touch on the importance of my own experiences that led me to the journey of investigating the psychological impact clinicians may experience when treating FGM survivors.

Theme's emerged that give some guidance to all professionals such as social workers, nurses and medical doctors of issues all need to be aware when working with FGM survivors, not just psychological practitioners. The book is composed of clinical work, through the narratives of seven therapeutic practitioners, the generalizability of this book is based on meaning making extracts. This book is a guide to help in the treatment of FGM survivors.

Types and medical consequences of FGM

The term 'female genital mutilation' refers to procedures involving the partial or total removal of the external female genitalia for non-medical reasons. There are four main types (see Table 1; World Health Organisation, 2012). An assortment of instruments is used to perform the procedure, which include knives, glass, razor blades and scissors (Al-krenawi & Wiesel-Lev, 1999); moreover, the practice is increasingly becoming 'medicalised', with doctors and other health professionals now performing FGM/C (Pearce & Bewley, 2014; Serour, 2013; Shell-Duncan, 2001).

The cultural position is complex: positive beliefs about FGM include preservation of virginity, improved marriage prospects, improved family reputation, or marking the passage into adulthood (Ahmadu, 2000; Boyle, 2002; Behrendt & Moritz, 2005; De Lucas, 2004; Keizer, 2003; Nienhuis, Hendriks, & Naleie, 2008; Yoder et al., 1999).

Table 1.1: Types of Female Genital Mutilation

Type	
Type i: Clitoridectomy	Partial or total removal of the clitoris and sometimes the prepuce (the fold of skin surrounding the clitoris)
Type ii: Excision	Excision: partial or total removal of the clitoris and the labia minora, with or without excision of the labia majora (the labia are the 'lips' that surround the vagina).
Type iii: Infibulation	Infibulation: narrowing of the vaginal opening through the creation of a covering seal. The seal is formed by cutting and repositioning the inner and sometimes outer labia, with or without removal of the clitoris.
Type iv: Other	All other harmful procedures to the female genitalia for non-medical purposes, such as pricking, piercing, incising, and cauterisation, or pulling and stretching the labia and clitoris.

Reasons given for the practice

Research outlines a myriad of explanations for why this act is committed. The most common reasons are, firstly, the controlling of women's sexuality, ensuring marital fidelity and preventing behaviour by women considered immoral, social pressure to conform and cultural identity/social cohesion (Boyle, 2002; Momoh, 2001). Secondly, rationalisations concerning hygiene purposes and preserving the virginity of women and girls (Lockhart, 2004; Yoder et al., 1999). Thirdly, a mistaken belief that it is a religious obligation (Behrendt & Moritz, 2005).

Fourthly, as a result of social pressure, girls may want to undergo FGM, or family members may often feel inclined to have this act done rather than feel outcast as members of their society (WHO, 2016). Research has also alluded to the procedure being an honour, or in some cultures, a transition to womanhood (NSPCC, 2017). In summary, the practice of FGM is deemed to be perpetuated for reasons to do with tradition, religion, psychosexual factors, sociological factors, and for hygienic and aesthetic purposes (Momoh, 2005; see Figure 1.2).

Figure 1.2: Elements of rationale for the practice of FGM (taken from Momoh, 2005).

Psychosexual	Religious
• A woman's virginity is an absolute prerequisite for marriage. • The clitoris is an aggressive organ that threatens the male penis. • Presence of the clitoris is central to sexual desire; its excision protects a woman from her over-sexed nature and saves her from temptation and disgrace whilst preserving her virtue. • In Egypt and Sudan, female circumcision is seen to increase male sexual pleasure.	• Clitoridectomy is believed to have its origins in African institutions and to have been adopted by Islam at the conquest of Egypt in 742 AD. Note: Female circumcision transcends religious boundaries. Female circumcision is not practiced in most Islamic countries and is not in accordance with the Holy Quran.
Social	**Aesthetic**
• An initiation rite of development into adulthood, becoming a mature woman. • Socialisation of female fertility. • The ceremony that often surrounds female circumcision is intended to teach young girls about their duties as wives and mothers.	• In some societies, the clitoris is considered unpleasant and ugly to sight and touch. • Removal of the 'unsightly' female genitalia is deemed a sign of maturity. • Female circumcision is thought to maintain the woman's physical and mental health.

Chapter II

The Global Impact of FGM

Prevalence and descriptions of female genital mutilation.

Female Genital Mutilation (FGM)[1] is a worldwide phenomenon, recognised internationally as a violation of the human rights of girls and women (United Nations Children's Fund (UNICEF), 2018). While the history of FGM is not well-known, the practice dates back at least 2000 years. It is not known when or where the tradition of FGM originated. It is believed that it was practiced in ancient Egypt as a sign of distinction among the aristocracy (Bilotti, 2000). Some believe it started during the slave trade when black slave women entered ancient Arab societies, while some believe FGM began with the arrival of Islam in some parts of sub-Saharan

---

[1] Until the 1980s, FGM was widely known as 'female circumcision' (FC) or 'female genital cutting' (FGC). The term FGM is used in this research, as it is widely prescribed in English-speaking Western societies. In April 1997 the WHO, United Nations Children's Fund (UNICEF) and United Nations Population Fund (UNFPA) issued a joint statement using 'FGM' rather than any other alternative

Africa. Others believe the practice developed independently among certain ethnic groups in sub-Saharan Africa as part of puberty rites. Overall, throughout history, it was believed that FGM would ensure women's virginity and a reduction in female desire (FGM National Clinical Group, 2018). The Romans performed a technique involving the slipping of rings through the labia majora of female slaves to prevent them from becoming pregnant, and the Skoptsysect in Russia performed FGM to ensure virginity. FGM is rooted in culture, and some believe it is done for religious reasons, but it has not been confined to a particular culture or religion. FGM is not mentioned in either the Quran or Sunnah (Lockhart, 2004). It has been highlighted that FGM was practised in the United Kingdom and United States by gynaecologists to cure women of so-called "female weakness" (Momoh, 2005). The practice of FGM continues in some communities in various forms today. Even in the 21st century, girls and women are still subjected to this harmful tradition (Gruenbaum, 2001).

The NHS website states that communities at particular risk of FGM in the UK originate from Egypt, Eritrea, Ethiopia,

Gambia, Guinea, Indonesia, Ivory Coast, Kenya, Liberia, Malaysia, Mali, Nigeria, Sierra Leone, Somalia, Sudan and Yemen. In countries that openly practice FGM, the incidence can be very high – such as in Egypt (91%) and Somalia (97%), even though these countries have recently declared FGM illegal. Female genital mutilation is a global problem that requires a global solution (Plan UK, 2015). Even in countries where FGM is banned, girls can be equally at risk, as the practice is often hidden. In addition, laws are often not effectively enforced and prosecutions are rarely sought.

Around the globe, 130 million girls and women have undergone FGM (World Health Organisation (WHO), 2014). In Africa, 101 million girls aged 10 and over have been subjected to FGM (WHO, 2014). Every year a further 3 million girls are at risk of FGM in Africa alone (WHO, 2014), amounting to almost 5,500 each day (Cook, Dickens & Fathalla, 2002). In the UK, it is estimated that 65,000 girls are at risk each year (WHO, 2014), and that one woman dies every 10 minutes from sequelae of the procedure

(Craft, 1997). FGM has been described as a specific form of child abuse; the perpetrators are usually family members or primary caregivers who have this done for reasons of cultural tradition (Toubia, 1994). Sixty thousand girls aged 0-14 were born in England and Wales to mothers who had undergone FGM. Approximately 103,000 women aged 15-49 and approximately 24,000 women aged 50 and over who have migrated to England and Wales are living with the consequences of FGM.

In addition, approximately 10,000 girls aged under 15 who have migrated to England and Wales are likely to have undergone FGM.

The UK impact of female genital mutilation

The Female Genital Mutilation (FGM) Enhanced Dataset (SCCI 2026) supports the Department of Health's FGM Prevention Programme by presenting a national picture of the prevalence of FGM in England. There were 1,630 individual women and girls who had an attendance where FGM was identified or a procedure related to FGM was undertaken in the period July 2018 to September 2018. These accounted for 2,025 attendances reported at NHS trusts and GP practices where FGM was identified or a procedure related to FGM was undertaken (NHS, Digital). There were 925 newly recorded women and girls in the period July 2018 to September 2018. Newly recorded means this is the first time they have appeared in this dataset. It does not indicate how recently the FGM was undertaken, nor does it mean that this is the woman or girl's first attendance for FGM. Between April 2015 and December 2017, 15,390 patients with FGM were treated in the NHS. Where we know what type of FGM patients treated between April 2016 and March 2017 had, there was highest incidence of Types 1 and 2 (35 and 31 per cent respectively). The average age of a patient with FGM, when presented to receive treatment in an NHS setting for the first time was 31 (between April 2016 – March 2017). Over 50 per cent of patients treated in the NHS were treated and live outside of London (between April 2016 – March 2017). More than 200 million girls and women alive today have been cut in the 30 countries in Africa, Middle East and south east Asia where FGM is continues to be practiced. Across the world, FGM is mostly carried out on young girls sometime between infancy and age 15. FGM is a violation of the human rights of girls and women.

The prevalence rate in the UK is growing, and FGM is a concern to all healthcare professionals including doctors, nurses, social workers, psychologists and many more (Rashid & Rashid, 2007). FGM was made illegal in the UK in 1985 (Prohibition of Female Circumcision Act 1985, currently amended to Female Genital Mutilation Act 2003) in England, Wales and Northern Ireland, raising the maximum penalty from five to 14 years in prison. However, research in the UK tends to be focused around physical rather than psychological health problems. There are medical procedures that reverse the effects of FGM (Erian and Goh,1995; McCaffery, 1995; Safari, 2013). This book will only focus on the possible interpretation of the psychological treatment for FGM survivors.

Chapter III

Perspectives on FGM

Researchers have considered many perspectives on FGM, including the cosmetic surgery debate, notions of patriarchal control, constructing the clitoris as bad, and human rights arguments (Jones, 2012).

Due to the varied nature of these perspectives, reflections in the book will specifically be made from the feminist perspective, the cultural relativist view and the standpoint of medicalising FGM. Another rationale pertains to the psychological and social perspectives of women experiencing shame if the act is not performed (Hayes, 1975; Hosken, 1993). "It is still claimed by men, that female sexuality is very dangerous and has to be controlled" (Hosken, 1993, p. 124).

Feminist: The feminist perspective advocates the notion of FGM being amongst the tools of patriarchal oppression. Penn and Nardos (2003) suggest that it is the belief that "powerful female sexuality" is a threat to social control that has led to extreme measures, such as FGM, being employed to bring about control and preserve the honour of women and their families. These assumptions in relation to women and the need to control them have resulted in the social functions of FGM (e.g. maintenance of chastity, attenuation of female sexual desire) being prioritised over the health complications that are often consequences of the practice (McNamara, 2002).

Cultural Relativist: The collectivist ethos maintains the cultural view that FGM should be accepted. This creates the false hope that undergoing the operation somehow improves people's lives—and the lives of their children—whether in the context of social status or of treating a medical condition, while the true reasons for the practice may lie elsewhere (Hellsten, 2004). Whereas, the individualistic ethos may be attributed to the Western ideological perspective of FGM being a violation of an individual's human rights (Gruenbaum, 2001). The emic perspective of FGM is bounded by the cultural embeddedness of this act, as it postulates that FGM is an act committed partly to accord with cultural standards and abide by the rules set through these cultural norms (Ahan, 2012; Gidden, 1993; Salmon, 1997). The etic perspectives, meanwhile, consider FGM as a practice as a cross-cultural dynamic and allow perspectives from Western culture, which establish FGM as causing serious bodily harm to the individual, contingent with the individualistic perspective (Baron & Denmark, 2006).

Some research claims that applying Western concepts of feminism, sexuality and human rights ignores the cultural values of the societies that participate in FGM (Adams, 2004). Views in the current media has been denoted as Western feminist evangelism (Okome, 2006). Beckett and Macey (2001) have argued that the division between discourses on human rights and the efforts to embrace "multiculturalism" in Britain has resulted in the acceptance of oppressive practices such as FGM. They concluded that "multiculturalism" not only exacerbates and legitimises the oppression of already-oppressed minority groups, but also poses threats to liberal democracy and individual human rights. Reporting anxiety about imperialism and concerns about who should involve themselves in discussions about FGM have for a long time resulted in the continuation of the practice non-FGM-practicing societies" to connote that while some communities within the broader society do practice FGM, the society as a whole does not (Jones, 2012).

For example: "In Africa, FGM is known to be practiced among certain communities in 29 countries: Benin, Burkina Faso, Cameroon, Central African Republic, Chad, Cote d'Ivoire, Democratic Republic of Congo, Djibouti, Egypt, Eritrea, Ethiopia, Gambia, Ghana, Guinea, Guinea-Bissau, Kenya, Liberia, Mali, Mauritania, Niger, Nigeria, Senegal, Sierra Leone, Somalia, Sudan, Tanzania, Togo, Uganda and Zambia. Certain ethnic groups in Asian countries practice FGM, including in communities in India, Indonesia,

Malaysia, Pakistan and Sri Lanka. In the Middle East, the practice occurs in Oman, the United Arab Emirates and Yemen, as well as in Iraq, Iran, the State of Palestine and Israel. In Eastern Europe, recent info shows that certain communities are practicing FGM in Georgia and the Russian Federation. In South America, certain communities are known to practice FGM in Columbia, Ecuador, Panama and Peru. And in many Western countries, including Australia, Canada, New Zealand, the United States, the United Kingdom and various European countries, FGM is practiced among diaspora populations from areas where the practice is common. Even though FGM is illegal in some of these countries the act is still prevailing" (United Nations Population Fund, 2017).

Medicalising FGM: Shell-Duncan (2001) asks whether medicalising female circumcision through a "harm-reduction" approach might help to protect women's health. She suggests that by considering a wide range of alternatives to the procedure that are still viewed to be culturally acceptable (e.g. nicking of the clitoris with a blade), the risk to women might be minimised. This proposition has been put forward in response to the fact that FGM has continued despite legislation prohibiting the practice. Harm-reduction strategies aim to minimise the health hazards associated with risky behaviours (e.g. intravenous drug use) by encouraging safer alternatives, including but not limited to abstinence (Shell-Duncan, 2001). However, opponents of FGM fear that medicalisation might legitimise a practice that is not only dangerous, but also serves to subordinate women within patriarchal societies (Dorkeenoo, 1995).

Physical effects

The effects of FGM include physical and mental health risk factors. The physical harms include chronic pelvic infection, formation of cysts, vaginal obstruction and infertility (Balgogun, Hirayama, Warkiki, Koyanagi, & Mori, 2013). Some of the most significant health problems associated with FGM are faced by women during pregnancy and when giving birth (McCaffrey, Jankowska, & Gordon, 1995). In some cases, complications from FGM during pregnancy can result in death (Annas, 1996; WHO, 2014).

Psychological impacts

FGM is often performed on pre-pubescent and adolescent girls, usually without anaesthetic, and using instruments such as razor blades (McClearly, 2004). Case histories and personal accounts taken from women indicate that FGM is an extremely traumatic experience for girls and women, which stays with them for the rest of their lives (McClearly, 2004; Kelly & Hillard, 2005). Young women receiving psychological counselling in the UK report feelings of betrayal by parents, incompleteness, regret and anger (Lockhart, 2004).

There is increasing awareness of the severe psychological consequences of FGM for girls and women, which can become evident in mental health problems, but professionals are not usually trained to deal with them because the patients' needs are seen as specific to their trauma (McCaffery et al., 1995). However, in her research, Lockhart (2004) noted that some women did not present with FGM as an issue for which they sought psychological support, as they described themselves as not having been affected by the ordeal of FGM.

Results from research in practicing African communities suggest that women who have undergone FGM have the same levels of Post-Traumatic Stress Disorder (PTSD) as adults who have been subjected to early childhood abuse, and that the majority of the women (80 per cent) suffer from affective (mood) or anxiety disorders (Behrendt et al., 2005). The fact that FGM is 'culturally embedded' in a girl's or woman's community appears not to protect her against the development of PTSD and other psychiatric disorders (Al-Sabbagh, 1996). Relating back to the collectivist perspective, there may also be psychological benefits to FGM for some survivors, such as feeling part of a community, not feeling different or dirty; feeling worthy of marriage, and being accepted in their communities (Khazan, 2015). However, the research also seems to suggest that these cultural aspects do not protect against psychiatric disorders (Chung, 2016).

There is currently no statistical evidence to affirm the prevalence of women seeking therapy for the psychological effects of undergoing FGM.

There is potentially also a categorisation problem, in that FGM involves issues of culture, race, gender, nationality, crime and child abuse (UNICEF, 2014). This may make it difficult for a therapist to decide what aspects need most attention (though of course this may depend on the therapeutic perspective). The mutilation may have happened abroad, leaving the sufferer to experience the effects in their later life while living in the UK (Toubia,1994). Whatever the case, the complexities may make it very difficult for both therapists and clients to manage disclosure in an unambiguous way, or even to decide on what might be considered a desirable outcome of therapy.

There is limited research on the experience of either clients or therapists in the context of FGM. Research has looked at women's experiences of FGM during the postpartum period, childbirth and pregnancy (Lundberg & Gerezgiher, 2008). A 2004 study looked at attitudes and experiences related to the practice, viewing FGM as mainly a parental expectation in terms of marriage prospects (Morrison, Dirr, Elmi, Warsame, & Dirr, 2004). Relating to health care practitioners' experiences, the overwhelming majority of research pertains to midwives and their experiences of working with women who have undergone FGM or face FGM-related issues in childbirth (Widmark, Leval, Tisherman, & Ahlberg, 2002). This may be why FGM has not received as much attention in the therapeutic world as other forms of child abuse, since FGM has been perceived to be primarily a physical health issue (NHS Choices, 2017).

Studies of the experiences of mental health professionals, specifically psychotherapists, are limited (Lockhart, 2004). The reasons for the limited nature of the research in this area are that FGM survivors, on the whole, do not see Western psychotherapeutic services as accessible or acceptable (e.g., they may fear judgement/negative evaluation by those services); that they may feel ashamed going outside of their culture; that they perhaps do not want to revisit the trauma; or even perhaps that, from their perspective, the West condemns their heritage and religion and they find this offensive (Chung, 2016).

The existence of specific guidelines is limited, and any existing guidelines for working psychologically with those who have suffered FGM are often mixed in with other areas that are arguably not compatible. For example, guidelines in national policy circles differ from multi-agency guidelines on FGM (Brown & Hemmings, 2013). Health and social care professionals in host countries—for example, in Europe, Australia, Canada, the USA and the UK—are increasingly encountering this vulnerable client group in their practice and are finding that they are ill-prepared to deal with the associated complex health needs and challenges (Momoh, 2005). In summary, as only limited research exists on the subject, there is a need for further research specifically on the experiences of both therapists and clients working in the context of FGM.

Chapter IV

Working with Trauma in Counselling

Issues in the therapeutic field when working with Trauma

In working with a client experiencing trauma, the therapist is challenged to meet the client in the pain of their experiences (Mearns & Cooper, 2013). Trauma has long been considered to be a transition, but it is even more existentially shocking. An event has occurred, often suddenly, which not only creates loss and the demand for reconstruction, but also fundamentally disrupts the whole assumptive frame upon which our sense of self is founded. The person's assumptions about the world, other people and their senses of self may be violently contradicted by one sudden catastrophic event (Mearns, 1977). FGM has been categorised by many researchers as a form of trauma to women (APA, 2015; Lockhart, 2004; WHO, 2008; Shell-Duncan, 2001). Six studies (Applebaum et al., 2008; Behrendt & Moritz, 2005; Chibber et al., 2011; Kizilhan, 2011; Nnodum, 2002; Vloeberghs et al., 2012) provided data on the prevalence of PTSD for women who have undergone FGM/C.

According to the ICD-10, trauma can be delineated within post-traumatic disorder. Trauma arises as a delayed or protracted response to a stressful event or situation (of either brief or long duration) of an exceptionally threatening or catastrophic nature, which is likely to cause pervasive distress in almost anyone (ICD-10, WHO, 1996). The traumatic nature of FGM can be shown by FGM clients having vivid memories of their FGM and coping with their symptoms in an avoidant way (Knipscheer, Vloeberghs, Kwaak, & Muijsenbergh, 2015).

In 2012, Vloeberghs et al. (2012) examined the consequences of FGM in the Netherlands. The sample was 66 women who had migrated from Somalia, Sudan, Eritrea, Ethiopia and Sierra Leone. One-sixth suffered from PTSD and a third had symptoms related to depression and anxiety (p. 677). All the women in the study reported some negative effects of stress, such as recurrent bad memories and nightmares. Vloeberghs et al. (2012, pp. 689-690) classified the women into three types. The first are the 'Traumatised': these women have suffered a lot of pain and sadness. They have recurrent traumatic memories, sleep problems and chronic stress; they feel misunderstood in their immediate environment and by health providers. These women may isolate themselves and experience a high incidence of anxiety/depression.

The second are the 'Adaptives': these women are overcoming the FGM experience and are able to talk about their concerns. Lastly, the 'Disempowered': these women are angry and defeated and do not talk about what was done to them, feeling ashamed, alone and disempowered. Thus, it is important to consider FGM treatment as a form of trauma therapy.

Figure 2: Core challenges experienced by therapists in working with complex trauma

Core challenges for professionals (Sanderson, 2013)

- Adequate training

- Knowledge and understanding of nature and dynamics of complex traumas

- Enough time to read and engage in CPD

- Enough time to reflect

- Awareness of power dynamics, both client and counsellor

- High-level of self-awareness – own abuse, especially attachment, shame re. sexuality, gender

- Feeling comfortable talking about trauma, sexuality and shame

- Acquisition and mastery of additional skills to manage the work without becoming overwhelmed

- Cultural sensitivity

- Socio-political context

- Integrating trauma therapy with existing therapeutic approach

- (Dis)courtesy stigmatisation

- Tolerating and managing uncertainty

- Understanding impact of vicarious traumatisation and secondary traumatic stress

- Counsellor self-care

Work with traumatised patients can alter psychotherapists' views of the world and of themselves, and can affect many aspects of their psychotherapeutic efforts. Being a therapist who works with trauma provides its own challenges. Figure 2 outlines the therapeutic challenges associated with trauma work.

Aspects in trauma therapy

Vicarious trauma, empathetic stress and compassion fatigue create the potential for therapists to be psychologically harmed by doing trauma work (Hernandez, Engstorm, & Gangsei, 2010). Research has alluded to the effects on therapists of working with trauma. One potential effect is 'Vicarious Traumatisation' (VT); this is defined by Pearlman and Saakvitne (1995, p. 31), as the "negative effects of caring about and caring for others". VT is the "cumulative transformation in the inner experience of the therapist that comes about as a result of empathic engagement with the client's traumatic material"(Pearlman and Saakvitne,1995, p. 31). Empathy is the helper's greatest asset, and also possibly his/her greatest liability. VT is not the same as burnout, although burnout may be exacerbated by VT. VT places emphasis on changes in meanings, beliefs, schemas and adaptation (Pearlman & Mac Ian, 1995). VT is more likely to lead to intrusive imagery and sensory reactions. Hatfield, Cacioppo and Rapson (1994) describe the type of emotional contagion that may lead psychotherapists to the "catching of [their clients'] emotions". VT permanently transforms helpers' sense of self and their world. VT can also influence countertransference responses. Harrison and Westwood (2009) identified some protective factors that mitigate risks of VT among mental health therapists, such as healthy relationships with significant others.

Features of therapists' response to working in the field can include, in particular, burnout, secondary traumatic stress, resilience and countertransference (Sanderson, 2013).

Burnout, which is often defined as a prolonged response to chronic emotional and interpersonal stressors on the job, consists of three components: exhaustion, depersonalisation (defined as disengagement or detachment from the world around you) and diminished feelings of self-efficacy in the workplace. It reflects a form of "energy depletion", and is often reported by counsellors when working with traumatic experiences (Maslach, 1982).

Figure 3: Common burnout processes

**Common phases in the burnout process
(Adapted from Freudenberger & North, 2006)**

- Compulsion to prove oneself
- Working harder
- Neglecting one's own needs
- Displacement of conflicts
- Changes in value systems
- Denial of emerging problems
- Withdrawal and isolation
- Changes in behaviour
- Depersonalisation
- Inner emptiness
- Depression
- Burnout syndrome

Secondary Traumatic Stress (STS) or Compassion Fatigue (cf; Figley, 1995) refers to the adverse reactions of helpers who seek to aid trauma survivors. STS is often used interchangeably with VT, although VT implies more permanent than temporary stress responses (Stamm, 1999). Common phases in the burnout process are illustrated in Figure 3. Figley (2002) explains how therapists' compassion is related to the ability to not only be empathetic but also to display an actual empathic response. This response is mediated by the emotional impact interaction with clients has on the therapist's concern for clients. The compassionate quality of therapists' responses thus depends on their perception of achievement through providing help, as well as their means of handling the stress inherent in the therapy process. In sum, both VT and CF highlight the negative impact of doing trauma work. VT emphasises the notion of cumulative stress and the inner experience of the therapist, while CF explains the factors in developing fatigue (Hernandez, Engstorm, & Gangsei, 2010).

In addition, Huggard (2003) developed research regarding STS 'doctors at risk, this study found issues relating to clinicians issues with empathy and detachment, issues in enhancing therapeutic relationship and issues in countertransference.

Concluding the key to effective management of STS is developing skills that enable early recognition of, and insight into, the strong emotions experienced within patient-clinician relationship. According to Huggard (2003) this leads to a possibility to embark on supportive and explorative processes that help establish appropriate responses to these emotions. Reflections of peer/professional supervision were valued. While a study by Meier, Back, & Morrison (2001) when working with trauma patients identified a range of emotions experienced by clinicians when responding to patient's needs. These included need to rescue the patient, feelings of failure, frustration and powerlessness; grief; and a desire to separate from and avoid patients in order to escape their feelings.

Resilience is a framework pertaining to psychotherapies that involves identifying and nurturing clients' strengths, fostering authentic relationships, and promoting personal control. "Mental health professionals cannot heal all the wounds suffered in tragic loss and humanitarian crises. What we can do is create safe haven to share both deep pain and positive strivings. Of value is our compassionate witnessing… for their suffering and struggle and our admiration for their strengths and endurance. We can encourage their mutual support and active strategies to meet their challenges. We can rekindle their hopes and dreams for a better future, support their best efforts and actions, and mobilise resources toward their aims" (Walsh, 2007, p. 244).

> **Possible presentation issues among trauma clients (Adapted from Muller, 2010)**
>
> - Anxiety about trusting the therapist, fear of dependency
> - Self-criticism, self-judgement, fear of being or appearing weak
> - Clients' use of distancing maneuvers:
> - An act of self-protection
> - A therapeutic opportunity
> - Issues with experiencing vulnerabilities
> - Client disclosure usually leads to feelings of being awkward, foolish and exposed
> - Based on attachment, the client, when feeling comfortable in the relationship, may present with feelings of:
> - Distancing themselves from others
> - Becoming disparaging (Bartz & Lydon, 2006)
> - Becoming unsympathetic (Haji, McGregor, & Kocalar, 2005).
> - Becoming aggressive (Logue, 2006).
> - Issues with connection

The impact of trauma in the therapeutic space and the therapeutic relationship. The role of attachment may be considered as an issue within the therapeutic space and relationship. Complex trauma within attachment can give rise to a range of attachment and relational difficulties. Many of these are a result of aversive early childhood experiences (e.g. FGM) and the structural dissociation that enables the survivor to 'split off' abusive experiences so that they can remain connected to the caregiver on whom they are reliant for care and protection (Sanderson, 2012).

Early experiences of caregiving give rise to templates, or inner working models (IWM) (Bowlby, 1973) of what can and cannot be expected in relationships and how others respond to needs. In complex trauma, children are not able to achieve self-integration or develop IWMs necessary to feel safe around others, which leads to disruption in the attachment system. As adults, this gives rise to insecure attachments in which they become either excessively dependent on others or fiercely independent. Shaping relational worth influences all relationships through the lifespan. Due to insecure attachment, relationships can become dangerous and terrifying rather than a source of pleasure or comfort, making them complicated and exhausting rather than soothing. This can be concomitant to issues of building trust within the therapeutic relationship and issues with the length of therapy clients seek. Presentation issues that may be experienced are described in Figure 4 .

Countertransference implies that the helper's response is influenced by the helper's own unresolved issues (e.g., lingering impact of the helper's victimisation experiences). This may lead to avoidance or over-identification with the client. The helper may take on a protective role for the client, becoming the client's "champion" of the client and adopting the role of "rescuer". The helper may inadvertently become a "surrogate frontal lobe" for the client, thus feeling the pain for the client. Consequently, therapists' experiences of understanding this specific form of child abuse may be helpful in identifying suitable self-care methods for therapist and healthcare providers; such as appropriate supervision, personal therapy and informed knowledge of this specific trauma. "Self-care is a skilful attitude that needs practice throughout the day" (Mahoney, 2003, p. 25).

Overidentification issues can occur when the therapist's empathetic response to a client is unconsciously intensified by his or her own abuse-related affects and cognitions. As a result of these unresolved issues, the therapist "overreacts" to those aspects of the client's situation most reminiscent of his or her own experience (Briere, 2002). In addition, overidentification can also be connected with cultural identity, as this may impact the work, either by the therapist's making assumptions regarding the client's experiences (and therefore hindering the therapeutic relationship) or sharing a similar cultural background with the client (which may promote a good working alliance within therapy) (Mishne, 2002).

Trauma education: Therefore, another aspect of trauma work is related to practitioner's education to the trauma work. This is related to the importance of training. In a study by Follette, Polunsy, and Milbeck (1994), 96% of mental health professionals reported education regarding trauma work (sexual abuse) was imperative to effective coping styles. In addition, Alpert and Paulason (1990) and among others (Chesterman, 1995) suggested that graduate programmes for mental health professionals need to incorporate training regarding the impact of clients' childhood trauma and its effects on VT,

Clients' trauma and therapists' understanding

Most recent research regarding working with FGM illustrates the importance of new literature that can enrich guidelines for helping with this client group (Abdulcadir, Rodriguez, & Say, 2015). Jones (2012) looked at clinical psychologists' experiences of working with FGM clients. In this study, 74 clinical psychologists working in a range of specialties completed a survey (quantitative approach). The findings indicated that participants had minimal experience of working with FGM-related difficulties. Knowledge about FGM and its consequences was also limited. Furthermore, clinical psychologists had received little training about FGM, and many did not feel confident in working with issues related to the practice. Implications for clinical practice and recommendations for further study include training opportunities specifically regarding FGM and future research exploring the connections between the physical and psychological consequences of the practice. In short, a counselling psychology perspective regarding the trauma of FGM is needed.

> **Three stages of multiculturalism in FGM survivors (Song, 2017), (*Figure 5*)**
> 1. Cross-cultural communication
> 2. Liberal egalitarianism and freedom from domination
> 3 Historical injustice and a postcolonial perspective

Although it would also be desirable to explore the experiences of clients, the ethical issues involved arguably prevent client experience from being a viable focus. Adams (2004) opines that research into the psychology of women who have undergone FGM is needed, suggesting that this research should ask questions about the counselling and support systems these women require. Cross-cultural research on the three stages of multiculturalism may also be relevant to understanding the lived experience of therapists working in this specialised area (Figure 5 outlines the three stages).

Therefore, a focus on therapist experience may be of particular help in gaining a purchase on such issues as cultural difference, moral ambivalence, and the complexities of dealing with an issue that challenges traditional categorical boundaries.

Chapter V

Pen Portraits

This chapter reflects on the participants involved by describing their experiences of working with FGM. The present study for this book used Interpretative Phenomenological Analysis (IPA) (Smith, 1996; Brocki et al., 2006; Smith, 1999; 2011; Smith, Flowers, & Larkin, 2009) to explore therapists' experiences of working with FGM clients. IPA's brief description is that it is an inductive qualitative approach to analysing data that allows for an in-depth exploration of the subjective experiences of individuals.

IPA involves the researcher taking on an active, dynamic and reflective role in making sense of individuals' experiences through interpretation (Smith, Jarman, & Osborn,

1999). This involves what is referred to as a double hermeneutic (Smith, 2008), whereby the researcher interprets an individual's interpretation of an event or experience. Smith and Osborn (2003) stated that, 'the participants are trying to make sense of their world, the researcher is trying to make sense of the participants trying to make sense of their world' (Smith & Osborn, 2003, p. 51). Reid, Flowers, and Larkin (2005, p. 23) further suggested that, 'IPA is particularly suited to research in unexplored territory', which makes it appropriate for an exploration of the psychological impact on therapists who have worked with clients who have experienced FGM. In addition, an IPA method is commonly used in investigations concerned with process (Smith & Osborn, 2003, p. 53).

Seven early to mid-career psychological therapists were interviewed about their experiences of working therapeutically with women who had experienced female genital mutilation. Three of the participants were recruited from charity organisations that work specifically with violence against women and had met with their clients through this organisation, while the other four participants had seen their clients within a NHS context. Each participant was given a pseudonym. Two of the therapists were recently qualified (within the last five years), whilst the other five were experienced, having been qualified for at least 10 years.

Pen Portraits

Angelina

Angelina had over 15 years of experience as a psychotherapist. At the time of the interview, she was a clinical lead, supervisor and manager of a women's charity. She had worked in the area of violence against women all her life, and FGM was something her charity had only recently received funding for. She was extremely passionate about her work with FGM. As a psychodynamic practitioner, I experienced her interview as one that was in-depth and 'deep'. I remember her recalling events regarding her experience that affected me. As my pilot participant, she uncovered many issues related to working with clients who had experienced female genital mutilation. Angelina had a personal resonance with FGM, as she was a white-British female married to a man from a culture that practices FGM. She uses her personal experience and her dedication to help as many women as possible. However, an important aspect of her role as a clinical lead focused around the care she took when assigning FGM clients to the most suitable therapist, as she states that this work is for women with experience in trauma work; this is because this issue does impact a person in an intrusive way, even if (like her) the therapist has many years of experience.

Catherine

Catherine had over 20 years of experience, and was a known psychotherapist working within NHS community settings. While her chosen modality was psychodynamic, she also found that being quite holistic in her approach to FGM clients suited the work better. I experienced her as quite reserved during interviewing; there were times she mentioned that she did not feel comfortable answering some of the questions because of the associated political connotations. It was difficult to explore what she thought of FGM generally. She often focused on the trauma aspect of FGM work and the emotional impact it has on therapists. Catherine was the only one of the seven participants to mention that one of the possible reasons for women to come forward for therapy regarding FGM was seeking asylum. Nevertheless, work with FGM clients was for her one of the most rewarding aspects of her job.

Anisa

Anisa was newly qualified with under 5 years of experience. She was a British- born Asian who experienced FGM work as one of the most difficult experiences she had had to date. She was a humanistic therapist, but at times would use eclectic methods in her approach. Her experience focused around one client that she worked with in a NHS hospital setting. She often described her guilt and fear surrounding working with FGM clients, as the media/literature at times had the connotation of Islam being the religion behind the practice. She often felt judged or felt shame associated with her feelings around her work with FGM clients, describing the work as overwhelming, and exploring her feelings of bracketing her own agenda to psycho-educate FGM clients that FGM is not an Islamic tradition. She often feared that because she wore a headscarf, this would affect the therapeutic alliance, with particular concerns regarding issues around transference; such as whether the FGM client might see her as either a mother or a sister. I experienced Anisa as open and defensive of herself as a Muslim practitioner. The impact of culture in her FGM work emerged as significant for her in her interview, whether it be her own background or the clients'.

Sarah

Sarah was a British-Asian clinical psychologist who had also researched FGM as a practice in her clinical training. Her interest began when she read a magazine that claimed FGM was 'part of the Islamic tradition'. She recalls that her activist side emerged. Sarah describes herself as integrative rather than having one specific modality. She describes FGM work as emotionally laden, although she empathises strongly with her client. She describes her feelings of empathy when, as she states, FGM 'was done out of love', stating that understanding the context around FGM was important in her work. Even though she stated that she believes FGM is wrong, she also explains from a cultural perspective that these women would have been cast out by society if they did not undergo the procedure, as it would have been difficult for them to get married and they would have been stigmatised and shamed. I experienced Sarah as honest and extremely understanding around the cultural dynamics of why FGM happens and the difficulty she has in calling it child abuse. She further compares Western ideas of 'designer vaginas' and how that is acceptable in this country, while issues around FGM are considered child abuse. She agreed to the issue of FGM not being an acceptable act, but was also extremely sensitive to client's experiences and their feelings about FGM. Context for Sarah was essential in her work in understanding and working with FGM.

Jennifer

Jennifer had up to five years of experience of working with FGM within a women's charity and the NHS. She considered herself as a humanistic therapist and an activist within her work with FGM. She also gave workshops to clients in addition to working with FGM clients therapeutically. She was British Somalian and spoke Somali with clients who found English difficult. I experienced her as very passionate about her work. Her interview focused around psycho-education as being essential in working with FGM. She describes some of her experiences with women comparing their vaginas and discussing which type of FGM was better, or, upon finding out that FGM was illegal, thinking that it would be okay if they opted for the lesser type. Issues around safeguarding emerged regarding the fact that women always need to be made aware of the seriousness of FGM and that there was no benefit of having FGM done to them. However, once she had completed therapy with clients it was extremely rewarding for her.

She mentions that therapists that intended to work with FGM clients need to be extremely open and understanding of the client's experiences. She states that, as practitioners, we need to understand that issues of trust may emerge. She also felt that women who come from a culture where FGM is practiced should be trained as counsellors and therapists, as language in her view is a key issue in the work that could be a hindrance; this is because some FGM clients may not trust therapists from other cultural backgrounds or find it difficult to articulate their feelings in English.

Elizabeth

Elizabeth was a clinical psychologist that had over 15 years of experience. Her experience of FGM work came through her employment at a sexual health clinic within the NHS. She was an Australian-Caucasian woman who described herself as a 'feminist', a broad concept, and working integratively. She describes her feelings around issues of safeguarding, the naming of child abuse, and how she feels t that more work around FGM is needed, especially in the world of psychology. She describes feelings of anger and sadness around the act itself, but also understands the contextual issues surrounding FGM. She further explicates that the necessary work does not have to be done in 50-minute sessions in a room, and that therapeutic help can be provided in many ways. She also compares the Western view to the cultural relativist view of issues around FGM. For her in particular, she states that she is lucky that the culture she is from does not commit FGM, otherwise she may have been a woman who had; it all depends on the culture you are from. I experienced Elizabeth as very open and vocal about her views. At times she would ask me questions such as 'what do you think?'; this was challenging during the interviews, as I did not want to hinder her meaning-making by voicing my view, and would consequently direct the topic back to the questions. For her, supervision was the more important factor in the work, as she described it a 'invaluable'.

Maria

Maria was a Black-British psychodynamic practitioner who had over ten years of experience. She was the last of the seven to be interviewed, and it was very hard to get her to take part. She described her scepticism of any research around FGM, as many people (in her opinion) were doing work on FGM for the fame or funding. Once she was able to read the information sheet and had a conversation with me, she was reassured that I (the researcher) was not one of these people, and that the research itself aimed to help future practitioners through learning about the experience of current practitioners. For Maria her interview focused around the horror of the act, as she described her vicarious trauma and spoke about not bracketing FGM as an act that happens in childhood. Maria had a variety of experiences through group and one-to-one work. She also describes that FGM can happen to anyone, as she had a female Caucasian British client who had this act done when she went on holiday with her husband who was from a practicing country. Moreover, a 20-year-old who got pregnant out of wedlock also had this act done to her. She focused on the impact of FGM, as well as on language issues Language was said to be important in work with FGM women, as the women described it in many ways: for example, Maria remembered one of the clients calling the woman who cut her a 'witch monster'. She often spoke of women speaking with their hands, describing things via imagery, and the importance of being aware that working with FGM clients can be hard.

Chapter VI

The Psychological Impact of Working with FGM

Vicarious traumatisation

The impact of the work led to thoughts about its psychological significance, as participants describe their lived experience of the trauma of listening to clients' narratives and how this has led to certain feelings that are related to particular facets of their work. In Maria's case, she named this as "vicarious traumatisation (VT)"; this experience of VT led to her feeling the need to do more for this client group. The impact was described as "needing" and "wanting" to do more for FGM survivors: "Yes, like a vicarious trauma actually, just being able to picture things that women described very, very graphically… I just felt like I really wanted to and really needed to do something more" (Maria, page 2, paragraph 10).

One of the therapists talks about her experience of secondary trauma by describing her client's trauma as something that is difficult to explore and feels 'present' in the therapy room. According to Angelina, this is a factor that leads women failing to return to therapy to talk about their FGM. She describes this using the analogy of the "rawness" of FGM for clients, which leads to working with individuals who are highly emotional: "It is a secondary trauma... If anything, the women that come for counselling in their FGM they are very much in still in their trauma and that is one of the difficult reasons why it is for women to attend as well. Because it is too raw." (Angelina, page 9-10, paragraph 40).

Angelina further describes her vicarious trauma as a physical feeling of being sick, of the many different reasons and when the act is performed the example she gives is one of form of punishment and then further worker with complications. So FGM is not always performed at a young age. The feeling of sickness is further attributed to the notion of the horror of the act, as well as being associated with the strong feeling of grief that she experiences as a therapist:

"I think there is an initial kind of physical reaction… like a real sickness… it comes a lot when I hear these stories, a real sickness… There is that kind of real sick physical of horror really of real horror which then turns into this kind of heavy grief." (Angelina, page 11, paragraph 46). Angelina describes her meaning-making of the psychological impact of FGM client work as VT, naming it as something she has experienced and suffered from: 'I think absolutely I suffer from secondary trauma' (Angelina, page 28, paragraph 112).

Anisa describes her traumatisation as being entangled with the emotions she experienced. Her exploration of these many different emotions leaves her feeling helpless, which is similar to something Maria mentioned in her experience of VT: namely, the need to do more. This can be seen as related to Anisa's feelings of being helpless because the trauma clients have experienced leads to practitioners' own experience of pain: "I think it… Because I begin to picture it, I think it does traumatise someone or it at least creates a lot of anger and sadness and helplessness, and I can really link with that" (Anisa, page 10, paragraph 94).

The symptom of vicarious traumatisation leads to the intrusiveness of trauma work with FGM clients. The meaning of intrusiveness attributed to FGM client work is unwanted, involuntary thoughts, images or unpleasant notions of FGM that can be obsessive, upsetting and distressing for the therapist. The work with FGM for all participants could be described as intrusive, since intrusive thoughts were something participants relayed as occurring when at home. In particular, Maria describes her thoughts of being more anxious regarding her own life: "I felt when I went home that night I suddenly became very aware of my own genitalia and of all the things that you think that you take for granted in life" (Maria, page 1, paragraph 4).

For one of the therapists, work with FGM clients impacted her personal life substantially, as she discussed thoughts about the sessions impacting her sex life in her marriage; these thoughts were described as unwanted and involuntary. The impact of the work affected her psychological wellbeing and her quality of life: "hmm but what I'm trying to say about that is how invasive it is and how difficult, it literally gets into my bed with me". (Angelina, page 6, paragraph 26). She described this notion of not being able to shake off involuntary thoughts ("One I think is as I said it is very hard to kind of shake off, I think it is something that I could get quite obsessive about...") as concomitant to the work, as her personal life was being affected: "I went through a period of time when I was having sex with my husband, suddenly these images would come into my mind of these women and what if you know... So it actually kind of got into bed with me...". She also supplied additional descriptions of feeling guilt as an emotion associated with her VT: "Because I'm thinking about the women who I have been working with has not got a clitoris and she cannot enjoy that" (Angelina, page 5, paragraph 20). The intrusiveness or the psychological impact of working with FGM was something attributed to an engrained experience that led to Angelina experiencing it as something difficult to let go of, even though she had years of experience: "...it's something that becomes very ingrained." (Angelina, page 4, paragraph 16).

One of the therapists described not being able to let go of the work, as well as the significant challenges she experiences when trying to have a normal life as the cases become intrusive: "However, it is challenging sometimes because sometimes you feel that, oh, you know, you're thinking about this woman's case..." (Jennifer, page 15, paragraph 180).

Sarah describes the psychological impact of trying to cope with the trauma of the act. As she relates it, it is intensely hard listen to the accounts of the survivors because of the level of trauma experienced by the practitioner: "It was difficult to kind of hear about you know... you know these are difficult things to hear about" (Sarah, page 15, paragraph 158).

As the literature describes, secondary trauma can be experienced in many ways. One aspect is the experience of physical sensations (McCann & Pearlman, 1999). Having bodily sensations associated with the work was something mentioned as being significant for Anisa, who described physical feelings that emerged when listening to the narratives of survivors: "I get all the tingly body feelings and even myself... As if it cuts my breathing; really just, like hyperventilating ..." (Anisa, paragraph 16, page 3). Anisa further describes her feelings of empathy with regards to her clients and states that she often empathises with her client's feelings, which impact her profoundly; there is a sense that Anisa experiences some forms of over-identification, as she has a similar religious affiliation to that of most cultures where FGM is performed. She further describes her empathetic experience regarding her frustrations and powerlessness associated with themes of human rights and familial issues of betrayal: "What if I was on that table and my own mom was holding me down?

And my sisters or siblings? And so it just really... Just a violation of the body and just the betrayal, family betrayal". (Anisa, page 3, paragraph 18).

Another aspect of the intrusiveness of the psychological impact of FGM work relates to the value of therapists acknowledging the need for their own self-care; if this is not established, some symptoms of secondary trauma stress may be experienced, as therapists will be more at risk of unwanted thoughts that could affect their personal life: "I think you can easily, take it home with you, worry constantly... think about it all night and, you know, be very, very worried about what's happened" (Jennifer, page 10, paragraph 127).

Burnout and exhaustion

For three of the psychological therapists, issues or symptoms of burnout emerged that could be associated with VT. For Jennifer, symptoms of burnout were exposed by issues relating to her having a similar cultural background to some of the clients. She described her feelings of finding the work tiring, over-identification with aspects of the experience that were akin to hers, and emotional exhaustion related to these similarities: "...speaking to women in the community who don't often talk about sexual health... feel like it's embarrassing and shameful...it has been challenging, it has been tiring... to be honest." (Jennifer, page 7, paragraph 77). Another therapist describes her burnout as exhaustion: "Well exhausted. I mean it's... It is exhausting, you know" (Catherine, page 2, paragraph 12). For Anisa, the feeling of burnout was assigned to her sense of being emotionally drained: "So it can be quite emotional. Emotionally draining" (Anisa, page 13, paragraph 122).

Working with FGM as a trauma

Some therapists had difficulty working with the trauma aspect of FGM. Through personal experience of working with FGM, some therapists believed the women that came forward for therapy did exhibit post-traumatic stress disorder or symptoms thereof: "The kinds of symptoms and difficulties these women have experienced, and a lot of the women that I've worked with have experienced post-traumatic stress disorder". Sarah, page 4, paragraph 34). Sarah further describes her concerns about survivors' PTSD and the lack of therapeutic intervention: "It's looking at what services are available for them, and to be honest there isn't anything, there is... Or there's very little" (Sarah, page 4, paragraph 34).

There are many attributions that therapists have delineated as the trauma aspect of FGM client work. This can be related to experiences of working with specific elements of trauma and what each therapist felt was the most impactful for them, such as experiencing clients' loss and grief because of this procedure. Catherine explores her construction of her most painful memories about the work: "But I think the most difficult times are when we see women that... I think when we see women who have been pregnant or who... Who are pregnant and who've lost babies, had miscarriages and... Or infertility" (Catherine, page 3, paragraph 22).

Anisa described her experience of listening to FGM narratives from a client as follows: "she'd be telling you this as if it was a psychotic episode" (Anisa, page 2, paragraph 12). In addition, she describes the client's detachment about the experience. There is a sense from her description that there is a depersonalised aspect to her client. The way she describes the client's engagement suggests the client is avoiding her emotional experience: "I don't know if it is as if she's transported back to that moment and so she blocks us all out and just starts, you know, quickly saying..." (Anisa, page 2, paragraph 12).

Some therapists specifically named FGM as trauma work, and therefore worked in a specific manner with FGM clients. Elizabeth, who understood the work as trauma work, therefore claimed it was fundamental to work with FGM patients and clients as though working with trauma: "I think we should be working with it as a kind of trauma. Because the idea of trauma is this is something awful that's happened to you, let's explore the effect that it had…" (Elizabeth, page 13, paragraph 80).

Maria states that for her, therapeutic modality does not matter so much in trauma work with FGM clients. Maria describes that within the context, particular therapeutic ethics needed to be adhered to. This can also be assigned to the ethics of relational trust. Furthermore, she denotes a more relational way of being in the work, which avoids the psychoanalytic approach of the 'blank screen': "…Just a slight moving away from that kind of whole analytic blank screen, not speaking, because it can feel very persecutory to the women who've suffered from profound trauma" (Maria, page 14, paragraph 120). Furthermore, Jennifer also relates an experience of working with a client's detachment like that mentioned by Anisa. Jennifer describes the issues of working with ambivalence to trauma with this client group as follows: "For me it's reminding myself that I will help a person.

Because some women who've undergone FGM will actually say that, no, I don't need any emotional support" (Jennifer, page 3, paragraph 160). This leads to Jennifer trying to manage her patients when working with the trauma elements of the work, describing the impact as having and learning tolerance for the work. She also describes elements of developing patience within the work, which may be interpreted as frustrations with FGM clients firstly not being able to engage in the work and secondly not being able to open up in therapy: "But it's about having the patience and I think I've had to learn that going along because..." (Jennifer, page 3, paragraph 160).

Clinical presentations were explored and attributions made to the trauma aspect of the work. One therapist mentions how the trauma of this work stays with her as an FGM therapist and the different ways this trauma is enacted: "it's important to understand how the trauma stays and how it gets re-enacted in different situations" (Angelina, page 15/16, paragraph 62). In addition, the parallel process with the clients regarding the trauma they experienced is also a characteristic of the work described by Angelina: "I think what I realise is there is parallels with other clients that I work with" (Angelina, page 15/16, paragraph 62). Correspondingly, issues with trust and boundaries are explored, as the predicament Angelia describes is focused around clients' early attachment bonds and the hindrances this may cause in the work, making reference to the lack of trust in the therapeutic relationship: "…if your very early experience is that you can be really hurt… It seems to be okay that your parents aren't stopping it, they're not protecting you". She also implies that clients may think they will be hurt by their therapist: "...so it's obviously accepted that you are a thing which is there to be hurt". Thus, Angelina describes working with the theme of "worthlessness"; she further describes the notion of 'repetition compulsion', a psychodynamic term referring to individuals choosing particular interpersonal relationships. In the case of Angelina's experiences, clients often choose relationships that are abusive: "...from an infant FGM survivors are in violent relationships or being abused in some way or the other, huge, nearly everybody on the whole has had some form of abuse, such as rape, childhood sexual abuse, domestic violence in relationships". (Angelina, page 15/16, paragraph 62). Her further description of shared feelings with clients of the act being awful, and her empathetic resonance: "They're here and they're traumatised, this is awful, why is this happening, and for me and I guess I share in that experience with the client thinking that this is awful, why has this happened to you". (Angelina, page 7, paragraph 30).

Parallels with client's helplessness

One therapist described parallel processes with the clients, reflecting her feelings of 'helplessness'. She attributes this to wanting to fix what has happened to her clients and somehow make it better. She further develops the notion of FGM being different to other forms of trauma work from her perspective, as it impacts on every part of the clients' being. She describes her feelings of anger and rage at the fact that these parents did not protect their children: "Well I think again as I said in the beginning there is a parallel with this sense of helplessness, you know how can I possibly fix this what I'm left with and I think, although our role is not to fix, there is this kind of feeling that you want to". (Angelina, page 14, paragraph 56)

Maria explores the psychodynamic concept of transference, describing the feelings of helplessness that she refers to as the client's helplessness that also lead to her feeling helpless. There is an element of being able to understand this as part of the work: "I think it's really important to look at the transference that happens between the… That transfers between you and the client… In the transference there is an overwhelming feeling of helplessness because that is being communicated" (Maria, page 2/3, paragraph 12). One therapist describes her feelings of helplessness at the injustice that has occurred to her clients. Further attributing issues of working with her client's fears of talking about their FGM, which leaves her feeling helpless as to how could she have helped: "This sense of injustice in the world and maybe the need to do something about it and that can actually leave me feeling quite helpless" (Angelina, page 8, paragraph 32).

Feelings of helplessness emerged for one therapist when considering womanhood and its implied possible loss: "But I think helplessness in terms of womanhood, I just really feel it". In addition, Maria explores this feeling of helplessness in terms of the psychosocial aspects of FGM. Helplessness was attributed to describing different forms of helplessness involving both psychological and social aspects of the work as imperative: "In a social, psychosocial way and in a psycho… In a, you know, psychopathology way and everything else. So, yes helplessness is a big part" (Maria, page 3, paragraph 14).

In addition, affiliations with available resources for therapists were also important to therapists' experiences, as the lack of resources available for FGM clients leads to one of the therapists feeling helpless, and a need for more help is raised: "We need to be plugging those resources and for me it was that feeling of helplessness…" (Sarah, page 5, paragraph 46).

Catherine describes her helplessness as a hopeless feeling, ascribing the notion of "damage" to her understanding and her experiences of the work, and defining the psychological impact for clients as an experience that leaves them feeling helpless; the damage is significant, resulting in a feeling of difficulty regarding the work: "…A sense of helplessness… the damage is done… The psychological damage I mean, it's… It's there for life." (Catherine, page 1, paragraph 79).

Therapist identity vs. therapist loss of self

A significant impact of the work with FGM clients was this subtheme of therapists identity that explicates parallels in experiences of the rationales therapist have heard for the reasons the act of FGM has been performed; such as the cultural/identity was an issue identified by therapists as being characteristic of work with FGM clients. Participants described many parallels with regards to therapists' lived experience of the sense of self, in particular their affiliation with being a therapeutic practitioner. Maria describes that mirroring the client's language and behaviour was one of the practitioner's identity issues she experienced, a loss of herself to help understand the FGM clients and their narratives: "Quite early on in the work I remember sort of the ladies waving their hands around, sort of in a general way or referring to genitalia as down there and that how in a kind of mirroring effect I was using the same type of language". Maria, page 1, paragraph 4). Mirroring clients and empathetically reflecting through body language is a way of being in the therapeutic space, although for Maria this was associated with feelings of trepidation: "...or I'd say, you know, and then I'd wave and then I kind of caught myself. I was like what are you doing, there's this mirroring thing and there's this fear of the unknown" (Maria, page 1, paragraph 4).

Catherine suggests that the feeling of loss in the work reflects the practitioners' feelings of loss within themselves, as the clients' feelings are dissociated and the same patterns happens for the therapist. This was another element denoting detachment from the work, as secondary trauma stress research reflects the coping mechanisms therapists employ to moderate their involvement with traumatic material. This can further can be seen as a consequence of FGM work: namely, disengagement and disengagement within the therapeutic relationship. This could potentially result in reduced levels of empathy: "You just feel the loss. You feel the loss and sometimes it's very difficult to stay present in the room because you can just feel almost... It's very bizarre. It's almost like you can feel the disassociation happening. Like you just can feel yourself withdrawing and to be very aware of all of that" (Maria, page 7, paragraph 48).

Another development of identity issues related to matters of therapists' own identity and cultural background emerged when practitioners worked with FGM clients: elements of the 'Western vs Eastern' cultural debate. This is also represented in Maria's lived experience. She discusses this by comparing her own treatment of her body in her personal life and the way FGM survivors describe their experiences of their body: "Well in terms of my own background I mean I suppose I couldn't come from a more opposite in terms of womanhood. I'm from Cuba, from a Latin American culture and just women we're just very encouraged to be very expressive and very... And it was quite shocking to... You know, one part of how subjugated they are as women within their community" (Maria, page 10, paragraph 74).

In reflecting on the dynamics of her own childhood upbringing and the differences in experience compared to this client group, Elizabeth describes her experiences of moments when she conformed to cultural norms, thus feeling some greater understanding of the work because of her experiences: "I mean I guess I'm aware that I had my ears pierced as a child because people did. I guess it made me thoughtful about things that I've been subject to..." (Elizabeth, page 7, paragraph 34). Comparing her culture to the FGM culture: "So, it's so different from anything that's been my own experience". There is a sense of relief in Elizabeth's narrative, as the culture she is associated with does not conform to such practices: "I think maybe being thankful that I grew up in a culture where women's views are valued" (Elizabeth, page 7, paragraph 34). In sum, Elizabeth's experience relates to other therapists' views, as they relate their own understandings of their culture the context of cultures FGM is practiced in.

There is an awareness among therapists of aiming to be empathetic towards survivors regarding their culture. Furthermore, Elizabeth distinguishes her own cultural identity and the paralleling of FGM survivors' experiences: "so many different complexities in terms of secrets within the family, cultural ideas, ideas of difference, being very aware of my white, middle class, feminist position, versus the different kinds of cultural groups, and positions that clients and patients have come from" (Elizabeth, page 1, paragraph 2).

Issues with the therapists' own cultural backgrounds affected both their work and their feelings around FGM as an act. Sarah illustrates this by discussing the associations made with regard to the Islamic tradition: "I know in my religion it's not an Islamic practice and it saddens me that those links have been made". (Sarah, page 3; paragraph 30).

Positive growth

Positive growth was a theme that developed for some therapists in their work. While descriptions of the work led participants to explore their initial difficulties, such as the VT mentioned above, working with such a topic and clients also resulted in an experience of therapeutic care with the extension of positive growth, establishing a feeling of being privileged to work with such a trauma. For Elizabeth, positive growth is connected with the notion of being privileged: "So, I think just the privilege of working with them, and, I guess, sort of, thinking what it took for us two women to be sitting and talking in a room about these things together" (Elizabeth, page 1, paragraph 2). Furthermore, Catherine also describes feeling as though FGM client work is a privilege: "I feel very privileged to do the work because it's..." (Catherine, page 3/4, paragraph 28). Angelina describes feelings of joy or personal growth when working with FGM clients. In addition to exploring issues of unresolved dynamics, whether it be client issues or clients not returning to therapy, there is an acceptance of unfinished and unresolved issues when it comes to this work: "It feels fantastic. That's why I'm here, that's why

I do what I do" (Angelina, page 26, paragraph 102). For Jennifer, positive growth was attributed to her experience of personal growth. This was distinguished by describing her experience as a psychological therapist, relating her experiences of FGM client work with forms of learning: "I was learning from them and it was quite surprising how open women were". There was also satisfaction, leaving her feeling a sense of accomplishment due to being able to help: "For me it was really rewarding to be able to give them the opportunity to talk. It made me feel, you know, helpful". (Jennifer, page 1-2, paragraph 11). Jennifer further explores how FGM work has caused her to grow as a practitioner: "for me personally as well it's made me grow, it's made me talk about topics that I wouldn't normally be talking about in your everyday life, and

I think it's been more of a benefit than a disadvantage for me" (Jennifer, page 9, paragraph, 111). Through her therapeutic work there was a parallel process with clients' feelings of being empowered: "That makes me feel empowered as well because I'm able to tell them that information" (Jennifer, page 13, paragraph 154).

Chapter VII

The Emotional Impact of Working with FGM

Feelings of sadness, heaviness and emotional pain were some of the emotions that emerged within therapists when working with FGM survivors. When Angelina describes her therapeutic work with FGM clients, she firstly discusses the compounded emotions: "There is a real heaviness, there is a real real sadness, that anybody had to experience FGM like that and there is kind of a pain, an emotional pain that really sticks with you it is very heavy".

There is a sense of being sensitive as practitioners, with descriptions of feeling sentimental about the FGM cases seen in practice ("...that we need to be sentimental about, that I can expect to listen to that and hear horrific stories about children being 7 and held down") and the process of helplessness that emerges in the work: "So sometimes I got to sit with the pain around it and hearing these stories and process it and somehow come to terms with my helplessness with it all". Furthermore, Angelina describes working with the past in the present by using 'inner child work' when working with FGM clients and their issues of suppression: "the work is great and we do lots of real a lot of inner child work. Hmm really looking back and healing that child, and thinking about what that would really have needed at that point". (Angelina, page 24, paragraph 98).

Feelings of guilt

Some of the meaning-making that was associated with participants' guilt centred around their identity as female practitioners and identifying with the loss of physical aspects of womanhood. Maria explores her guilt about having a fully-formed vagina and the issue of this possibly affecting her empathy: "I felt a lot of guilt. I felt guilty that I was sitting in this space with the women having a fully formed vagina and I was talking to them about them not having and they had no idea whether I'd had the procedure or not" (Maria, page 1-2, paragraph 6).

One therapist experienced a feeling of unconscious guilt regarding her own experience of receiving joy in sex, something she felt her clients could never experience. In making this attribution, she possibly felt both a sense of sympathy and a block to empathy, as it was something difficult for her to imagine: "...if the guilt was there it was more on the unconscious level... Is that I have this ability to experience joy and pleasure naturally and that they don't and what could that be like. What could that possibly be like, how do you manage to live without that ability to experience that much pleasure". (Angelina, page 6, paragraph 24).

Another feeling that emerged was religious or cultural guilt. For Anisa, being a British Muslim, this feeling resonated within her experiences of the work: "I felt as if it was my fault or, like, because it's my religion that it was done under, sort of thing. So it felt like" (Anisa, page 1, paragraph

6). For Sarah, the difficulty lay in hearing that FGM was a religious practice: "As a Muslim hearing people, hearing these women saying that they believed or they believe it's an Islamic practice. I found that very difficult". (Sarah, page 5, paragraph 48).

Feelings of sadness

The feeling of sadness engulfed all of the practitioners. For Maria, being a black female practitioner impacted her work: "Well I suppose as a black woman, as a black practitioner, it makes me feel quite sad" (Maria, page 9, paragraph 68). The emergence of sadness in the work can help with practitioners' ability to maintain their own self-care as therapists. Sad emotions led Catherine to be aware of the severity or the level of impact in the work: "Well I mean it's, you know, it's very moving and it's... The stories are very sad... That's why I, sort of, pace the work and take it slowly, even when we're very busy" (Catherine, page 3, paragraph 24).

At times sadness was associated with guilt; this was described by Anisa in three ways. Firstly, she describes it as guilt with regards to the need to do something for these women, concomitant to feelings of sadness: "Yes, and the feeling I need to make it up… just feeling really sad". This was followed by Anisa describing both her empathy and her sense of good fortune as she had not experienced FGM: "I thought how fortunate I am that it never happened to me because it could have been. So those two things" (Anisa, page 1, paragraph 8).

One therapist referenced her feelings of sadness akin to the sadness practitioners may experience when working with other forms of child abuse: "…makes you feel very sad, makes you feel… Yes it's just one of those things like child abuse…" (Sarah page 15, paragraph 164).

Concomitant to the feelings of sadness was the notion of loss. Maria emphasises the importance of working with loss in clinical work with FGM survivors: "never, ever underestimate the experience of loss in that". She expands on the theme of loss with a variety of detailed descriptions. The first concerns loss of childhood, followed by loss of hope, with descriptions of the FGM survivors' loss of a physical part of their body as being connected with loss of clients' identity, in particular loss of their womanhood: "There's layers and layers and layers and layers of the loss. Loss of childhood, loss of faith, loss of intimacy, loss of you know, the physical parts of yourself, you know, loss of womanhood. It's just a lot of losses, you know" (Maria, page 18, paragraph 167).

Loss emerged as a sub-theme for one of the therapists around specific issues she experienced when working with FGM clients. For Catherine, there is a notion of understanding FGM survivors' feelings of loss, connected to the lived experience of listening to clients' descriptions of their loss and not feeling that clients can come back from their loss: "this sense of loss. Strong sense of loss. They feel they can't recover from". Additionally, for Catherine personally, this led to her having an interest in post-ante-natal care when working with FGM survivors, given clients' loss of the ability to be a mother: "I'm particularly interested in post-natal women, ante-natal women personally" (Catherine, page 13, paragraph 121). Catherine further associates the connotations of FGM with feelings similar to a bereavement: "that sense of loss. So that, kind of, almost like bereavement feeling". Acknowledging the grief process associated within the work: "have been through a process of grief really" (Catherine, page 6, paragraph, 43). The word 'damage' resonates with Catherine, as her experience of this procedure for the women she sees in practice is this sense of damage: "Infertility I think is the most... Probably one, if not the most powerful kind of reminder of the damage" (Catherine, page 3, paragraph, 26).

Feelings of anger

For three of the seven therapists, the feeling of anger impacted their work. Anger was represented in many ways. For Maria, this emerged in the interview as feelings of discord with the work: "Yes, I feel really angry. Really, really angry. And sometimes I feel quite conflicted as well. Sometimes I think I've just got to work with the work". Furthermore, anger towards the act of FGM itself was interpreted as a conflict between her cultural and personal beliefs; this further highlighted the anger, since her anger was embedded in her perspective which she described as Western: "I don't want to feel like I'm pushing a Western perspective on things that are wrong in my eyes but culturally embedded" (Maria, page 10, paragraph 76).

For Angelina, a myriad of feelings of anger emerged because of the culture around FGM, the lack of education in these communities and the issue that the perpetrators of this act are 'the mothers' of the survivors: "My anger and rage is towards to, well it's a mixture. It is kind of towards the society that condones it, it is towards the actual cutters, who carry out these horrific operations don't know if you can call them that on children and young women and it is kind of deeper for that for me". The participant further described issues that are brought into therapy, such as difficulty in building relationships, as the primary care giver did not protect the client from this procedure/act. Angelina raised the issue of dealing with hurt but understanding the client's perspective that it was not done in a malicious way: "I say I am angry, it is difficult to be angry when I can say I understand as well, why people continue to do it and why they think it's okay to do it. So there is a very mixed and I think that reflects on the conflict that the survivors of FGM have as well" (Angelina, page 1-2, paragraph 6).

For Anisa, this feeling of outrage was intertwined with hurt specifically attributed to the act itself: "It angers and upsets me because of the association and the fact that it happens…" (Anisa, page 9, paragraph 79). Anisa explores her anger at the religious context of FGM, comparing it to her religion. The impact of her intense anger is connected to her own knowledge and personal resonance with religion: "I think that it impacts on the fact that it's under the guise of Islam and under the impression that that's… Is… Then that angers me because it's not!" (Anisa, page 4, paragraph 24).

Another feeling revealed via the sub-theme of anger was the feeling of frustration in work with FGM clients. This was explored through the need to educate FGM clients, struggles in the work, and lack of clients coming to therapy: "then frustration of wanting to educate her that" (Anisa, page 4, paragraph 28).

Jennifer explores her irritation regarding the different concerns regarding her problems that may arise in the work: "think it could be a little bit frustrating… that when they're ready they can talk to me. Might be six months down the line when they're ready to talk about it, it might be a year you know?" (Jennifer, page 14, paragraph 162).

Angelina describes her feeling of frustration around FGM and the fact that this act may be happening in the UK. She explores the different ways women get cut and her feelings associated with FGM being a worldwide problem passed down from generation to generation. She describes her fear that FGM occurs on such a large scale and yet therapy is not accessed; this leads to Angelina describing a need to do something, leaving a feeling of frustration: "there is a huge huge sense of frustration, in that what are we doing. It's not enough". She further acknowledges FGM work as engulfing a practitioner with emotions and thoughts: "how do you stop it from completely taking over and being your life's work really". Consequently, describing it as something difficult to let go of: "it is really invasive work, to shake off" (Angelina, page 27, paragraph 107).

Chapter VIII

Cultural Dynamics in the Work with FGM Clients

The figure above represents parallels with issues that practitioners described that clients experience. Explorations associated with ethical dilemmas emerged, such as child abuse, the context of the practice and issues with current legislation.

Cultural embeddedness

The cultural dynamics of the rationale regarding the practice of FGM were essential to all therapists' views regarding their work with clients. Jennifer explores her lived experience of seeing FGM as deep-rooted in the culture: "It's very culturally embedded, it's very deep-rooted…

it's quite astonishing how widespread". With reference to the potential rationale for the act in her experience: "it's patriarchy more than culture because a lot of the cultures are not even that similar". (Jennifer, page 16, paragraph 196).

One therapist had a feeling of despair attributed to the survivor's mothers, noting that she understood the social dilemma while on the other hand having anger towards them to allowing this act to happen to FGM survivors: "It is a kind of mixture of anger and despair really... How could you allow that to happen to your daughters? However, I always see the dilemma in a social context" (Angelina, page 3. paragraph 8).

One therapist spoke of her dilemmas when working with groups, the many different groups within the different cultures practitioners need to be aware of, and the difficulty that may arise in the work. For Maria, this was described by the different sectors of FGM in addition to her lived experience of the age differences; "...Women from different parts of Africa and the differences in age, reason, type of

FGM in one group is quite remarkable how different it can be and all the problems that that in itself creates". The challenges that emerged in the work; in particular descriptions to the way this was experienced. Empathising the impact, she encountered regarding the challenges with the different demographics and experiences from FGM survivors. Some believed they were survivors of FGM, while others were disregarded; something that can be attributed to the continuum of the act: "in terms of if you had it later the ones who'd had it earlier felt that the ones who had it older… At an older age had no right to be there because they had had an experience of having a clitoris or they knew what it was like to exist before the cut" (Maria, page 4, paragraph 20).

Understanding the cultural aspects of the work led to the therapists stating their perspectives on the intentionality of FGM as an act and their strong views about the injustice that occurred: "Yes, unfair. I felt like it was a big injustice to women" (Jennifer, page 11, paragraph 132). An understanding of the practice is explored in Catherine's experiences with this client group. She describes this as understanding the context, to a certain degree a holistic view of FGM done out of the feeling of 'love'. This accentuates therapists' understanding of the clients' subjective world: "The women always say to me, not always, but mainly they say it was done out of love. Most of them have forgiven their mothers for it. What they want is to live with it as best they can and protect their children..." (Catherine, page 10, paragraph 81).

For Elizabeth, her cultural view explored her own identity, how she has been able to relate to clients and her ability to empathise with clients by self-resonance: "From a migrant position so what it's like to be in a new culture, and how that actually might be similar, in some ways, to my own experience because I was an Australian working in England". (Elizabeth, page 5, paragraph 24,). Elizabeth further explores the culture of psychology, her experience of causes, her being mindful of her own cultural policing, and her concerns with the aspects that are helpful regarding FGM client work and the unhelpful aspects with regards to the culture of psychology: "...I think it's holding in mind all of these different perspectives and being able to hold different views and to reflect on them and to think with women...being able to deconstruct ideas about how helpful is it to think about it as something that is culturally sanctioned that will help you get married, how helpful is it to thinking about it as something that's child abuse and you might do it to your children? (Elizabeth, page 6, paragraph 32).

One of the therapists feels that therapists need to have in-depth knowledge of the cultural embeddedness of FGM: "hmm what happens a lot and the cultural importance of it. I think it is really important to be informed about different cultures, the way different cultures practice". In addition, she mentions that practitioners need to be aware of the importance of self-care: "I think that's important. I think self-preservation is important and self-care is really important, although as much as you kind of look after yourself". (Angelina, page 15, paragraph 60).

Explorations were made regarding the dynamics of FGM, issues with culture and the difficulty therapists encounter regarding clients' views of their FGM. This was attributed to the lack of FGM clients seeking therapy and the fear clients experience further describing experiences about the clients accessing counselling; with descriptions of a therapeutic continuum among this client group, such as people drop out, not accessing therapy to individuals who therapy has been the best experience for them: "The counselling is not something that they're familiar with". On clients accessing and getting the help they need: "It can be, sort of, a little bit frightening. I mean on a continuum from absolutely not touch counselling under any circumstances, to just..." (Catherine, page 1, paragraph 4).

Due to the cultural elements that FGM is associated with, such as an Islamic practice and a societal patriarchal practice, most psychological therapists agree it would be culturally inappropriate for a male therapist to work with female genital mutilation survivors: "I just think culturally it wouldn't be appropriate for them to see... so for me I don't feel comfortable with it, but..." (Maria, page 17/18, paragraph 161).

Angelina explored her feelings concerning male therapists and how this was difficult for her to imagine, as FGM (according to her) is a secret act that should not be spoken about to the men from that culture; moreover, as there is something quite intimate about it, there is a sense that clients would feel uncomfortable. In her opinion, therefore, a male therapist would not be suited for work with FGM clients: "I think women would find it very difficult in my experience... FGM is performed in secret most of the time... I don't think that any women that I have worked with would be comfortable working with a man". (Angelina, page 19, paragraph 80).

Jennifer further explicates her views of male therapists working with FGM clients; her account relays a strong disagreement with the idea. Her meaning-making regarding this experience is justified by survivors describing their desires regarding the gender of their therapist: "I personally disagree with a male psychotherapist... I've had cases where women said that they don't want to be in a room with other men and talk about the subject and I completely understand that". (Jennifer, page 16, paragraph 200).

Catherine, however, outlines her thoughts on the positive impact a male therapist may have on FGM survivors. Firstly, Catherine describes her explorations regarding the psychological concepts involved: "I think the transference would be interesting". Secondly, she explains what she thinks may be helpful in terms of 'repairing the damage'; this can be attributed to the association of FGM being a culturally embedded procedure that is male-led or instigated by men: "How good would that be, you know to... How... How to repair, in a sense, the damage" (Catherine, page 12-13, paragraph 113).

Due to the cultural embeddedness of FGM, it was described by one of the clinical psychologists that having women from the same cultural background as the client seeking help would be more suitable, according to her subjective experience: "Professionals who themselves have come from those cultural backgrounds; have come from those religious backgrounds who can actually... Who strike a chord with these people" (Sarah, page 3, paragraph 30).

Language barriers and issues

Another important characteristic that emerged from the therapists' experiences was the issue of language. The use of specific terminology and the way language was constructed (such as the context in which descriptions of the act took place within participant experiences) were illustrated as important factors to help with understanding and coping with the work. For Maria, her experience largely focused on the language barriers and her experience of the different tribes. Describing FGM as very unique, she explores the use of fantastical language: "Like in Sierra Leone they're called Bundu devils, the women who do the cutting". She further explicates her own meaning-making by looking at this via Western and Eastern views of FGM. Comparing this to a dilemma in another context, it could be interpreted as a psychological diagnosis similar to symptoms of psychosis: "Looking through a Eurocentric lens about this fantastical language, in any other sense this kind of language would probably indicate some sort of psychosis" (Maria, page 4, paragraph 26). Nevertheless, Maria does explore the importance of understanding the survivor's subjective experience, as she describes the issues with language she experienced as clients made sense of their reality: "Their subjective world and that they'll use language that's maybe not familiar to everyday whatever theoretic practice you've been trained in". Further, acknowledging the importance of further work in this field and how to use language as a tool within FGM client work: "How to recognise that language or how to make sense of how they're making sense of their reality" (Maria, page 5, paragraph 28).

For Elizabeth, language can also be seen as an essential factor, both in the therapeutic work itself and due to its ability to manage expectations and help with understanding the dynamics associated with FGM: "I think also I am aware that, you know, the way that we use language is very important…You've got to manage expectation really clearly and talk about what we can do is" (Elizabeth, page 4, paragraph 22).

Tailoring therapy to the client's needs through language was another facet of language issues: "And I think you've got to tailor to the clients' needs. Just as you would in terms of language needs" (Jennifer, page 18, paragraph 227).

The use of language could also help future work. Focusing on how clients construct their experience is valuable in working with FGM. Elizabeth finds comfort in describing her experience of understanding the context of FGM through clients choosing whether or not to name their trauma as a mutilation. This was relayed by Elizabeth in describing another form of violence against women, using the concept and experience of "rape" and the way that practitioners listen and follow a client's story rather than pushing agendas: "In some way join with a patient's language, if a patient is saying he had sex with me when I was asleep and I didn't really consent, you wouldn't be pushing an agenda of, well, that's rape, that's definitely rape, if the patient didn't construct it as that". (Elizabeth, page 13, paragraph 82). She further describes her meaning-making of understanding language constructs as "sitting alongside the patient… their view" (Elizabeth, page 13, paragraph 82).

The need for more practitioners from cultures that practice FGM is evident if more clients are to be encouraged to come forward; similarly, as mentioned earlier, disengagement of some FGM survivors who do access therapy was an experience that emerged in the interviews. One contribution to the lack of engagement or survivors not seeking therapy is language barriers: "I don't want them to have a barrier like language that prevents them from getting support..." (Jennifer, page 2, paragraph 15).

Using terminology that is idiosyncratic to the client's narrative is important; there are elements of humanistic counselling values at the centre of the work: "we need to treat her with dignity and respect as a human being who's undergone a traumatic experience". Humanistically, the methods clients use to construct their trauma are connected to issues with language: "Does she still consider herself a victim or a survivor? The terms that she's comfortable with using as well it's important to acknowledge..." (Jennifer, page 5, paragraph 49). Sarah further describes the dilemma of her work with clients concerning whether or not to refer to

FGM as an abusive practice: "So I think I'm very sensitive about the terminology that I use. That doesn't mean that I don't believe it's a you know, an abusive practice..." (Sarah, page 16, paragraph 168).

Furthermore, Sarah voices her concerns with language used by practitioners. The rationale for the sensitivity she employs in her approach is the aim to engage FGM clients rather than disengaging them; in Sarah's experience, this is achieved by avoiding assumptions in the work: "If you're trying to work therapeutically with someone then obviously you need to use terms that are going to engage them, not terms that are going to alienate them and make them think, gosh this person's come here with assumptions" (Sarah, page 16, paragraph 170).

Child abuse vs context

Working with FGM leads to the paradoxical issue of viewing FGM as child abuse versus understanding FGM from its cultural context.

Elizabeth constructs her view on the dilemma as FGM as a different form of child abuse: "The word child abuse is framed in such a different way for every other act". It could be understood from her narrative that the perpetrators in these instances are, unusually, family members (in most cases the mothers), although there are also considerations regarding the procedure being perceived as being done out of love "I think there's something very different about… FGM occurs in loving families". While Elizabeth states that FGM should not happen to children, her position on naming the practice as child abuse is dubious: "As something that shouldn't happen to children, versus, or something awful or, you know, but I think the term child abuse is so in the culture of neglect and damage, and it's the intent" (Elizabeth, page 5, paragraph 28).

There is an appreciation, or a learnt understanding, of the contextual issues of this being a cultural paradox. Elizabeth recalls her dilemma in making sense of her experience of the FGM procedure, through being able to understand the reasoning: "well I am doing something that will help my child get married, not be ashamed, not have stigma about being loose, or unclean" (Elizabeth, page 2, paragraph 12). Using the words 'child abuse' when talking to clients, or describing FGM to clients as child abuse, can lead to issues of silencing the clients; again, Elizabeth describes the lack of FGM clients presenting for therapy, along with her experience of rationalising the disengagement in the work. Because of the term 'child abuse', according to Elizabeth: "But if you see in the papers that FGM is child abuse, you're not going to go into the clinic and go, well, I'm going to get my daughter done. I think also there is a silencing in the way that we position it too" (Elizabeth, page 12, paragraph 90).

For Sarah, it was valuable to understand and contextualise the act of FGM when working with FGM clients. Sarah described this as working with the presenting problem or issues encountered in therapy, as FGM may be a factor in the issues clients experience. Sarah further explores the importance of clients not wanting to work with their FGM procedure as their main focus in therapy: "you've got to contextualise... when actually when you put it in context for that particular individual, FGM might have been difficult but actually the fact that they saw family members killed in war is much more traumatic for them." (Sarah, page 1, bottom of paragraph 2).

The difficulty for one of the participants was in naming FGM as child abuse. According to her, as she felt she understood the context, she disagreed with the term: "I think for me the main issue that I've had with the whole way that FGM is being dealt with in this country is the way that it's kind of categorised as child abuse...". (Sarah, page 2, paragraph 20).

Catherine, who has difficulty in understanding FGM as child abuse, describes the Eastern vs. Western debate regarding the context of FGM procedure: "No, I'm not saying I don't agree with it but it's just that... You know, it is a... It is a Western concept, child abuse, and most of these women would say that this was done out of love for them" (Catherine, page 7, paragraph 51).

Issues with risk/ethics

Working with FGM in the UK means that legislative issues may arise. These concerns and experiences were important to acknowledge, as were the safeguarding rules and procedures. For one of the therapists, the important factor was working with issues of safeguarding, as she expressed a tentative approach to the work: "under the safe-guarding bracket you've got to deal with it very sensitively" (Sarah, page 18, paragraph 190).

The issues in clinical practice that are experienced by practitioners because of FGM being a safeguarding issue leads to clients feeling fearful about coming forward. This is one of the experiences described by clinicians: "it's all very kind of safe-guarded, child protection... It's got that kind of taint to it as well. I think that will put people off as well because they will be worrying that if I present to a service and mention that I'm struggling with this are my parents going to get prosecuted" (Sarah, page 5, paragraph 46).

One of the therapists expressed her feelings of relief regarding her experiences with women who have had FGM and are against their daughters being subjected to it: "'No! this is not going to happen to my child!' I'm breaking the cycle of abuse here. I am breaking this tradition and I am trying to educate other people. So absolutely yes I have had this, and hmm I breathe a sigh of relief when I hear that" (Angelina, page 3, paragraph 12).

Further to the subject of dilemmas experienced within the work, the fear of seeking therapy became a theme because of the tainted nature of the current legislative associations. Thus, issues with the current legislation surrounding FGM causes clients to feel reluctant to open up: "Especially that legislatively how to work with FGM, so what I've known is just that it's really silenced my clients…noticed that that areas the women don't talk about so much anymore and I think they're really frightened" (Maria, page 5, paragraph 32).

Ethical issues become particularly apparent when FGM is viewed as child abuse. Ethically, FGM was experienced regarding the way the procedure was positioned. Regarding the view of FGM as child abuse in terms of mothers being abusive, or the ethical ways to help women other than using therapeutic interventions: "Ethical issues…who is this a problem for, and why and who is it a problem not for?… how is the problem positioned… is the problem abusing mothers, and the solution is medical interventions?" (Elizabeth, page 8, paragraph 48).

One therapist describes her experiences regarding the dilemma of FGM being named as child abuse: "just when talking about that particular aspect of child abuse, we have women coming in increasing numbers who are claiming asylum" (Catherine, page 8, paragraph 53).

The risk issues that emerged for one of the therapists in the course of her work were that the naming of FGM as abuse alarmed survivors, which led to them not seeking appropriate care: "I think there's also real concerns that women wouldn't engage in ante-natal care, because there were concerns that if they were a survivor of FGM they would be positioned as a potential abuser and that social services would get involved" (Elizabeth, page 12, paragraph 72).

As repeatedly mentioned within the work, the issues of the shortage of women seeking therapeutic help was attributed to the law regarding FGM, a possible impingement to the work. Thus, making clients aware of the law caused issues: "It's hard because as part of my job I do have to tell them about the law...I have that duty to tell them about that this law is in place and that, you know, FGM is considered child abuse, it makes it harder to gain the client's trust, it makes it harder to work with them on a more intimate level, and to be able to have them trust you completely and tell you their experiences". (Jennifer, page 3, paragraph 22).

Jennifer recalls the many different presentations, describing the impact of seeing FGM clients from different cultures and age groups as an experience that made her aware that anyone could be at risk: "Yes, we've had a very kind of recent case is where we've had an Eastern European woman at risk of FGM because she's married into Western African man and his family"(Jennifer, page 19, paragraph 225).

It was implied that due to FGM's appearance or the way FGM is experienced by clinicians, because of the associations with risk, there is a sense of panic or the need to do something. However, this was seen as being done more abruptly in Sarah's experience, therefore having a negative effect in the therapeutic work and also highlighting the difficulty a therapist may experience if there is a risk. This may not be dealt with in a proactive way; perhaps a sensitive or different approach is needed when FGM is revealed as an issue: "The only way it comes to light is when professionals practitioners are realising that a child or whoever has been perhaps subjected or is at risk of being and that's the only time that it's coming to light, in a very reactive kind of way. I don't think there's anything proactive going on" (Sarah, page 4, paragraph 36).

Chapter IX

Therapeutic Implications of Working with FGM Survivors

The need for psychoeducation and important aspects of the work

Therapists explored their own concerns regarding their meaning-making of naming the need for psycho-education in the work as paramount. One therapist explains the importance of being 'well-informed' when working with FGM and its traumas. She feels that from personal experience and through her travels, she has more of a handle on the issues described by her clients: "Well I think it is really important to be really well informed and I can't profess to know everything about FGM of course but it is really important to be well informed". (Angelina, page 14-15, paragraph 60).

Anisa further explores feelings of needing to educate clients, making possible suggestions for future therapeutic work similar to the CBT modality of psycho-education; for clients in Anisa's narratives, the psycho-education is about the FGM procedure: "I think it just needs to be talked about a bit more and people need to educate about it a bit more and... maybe not everyone but especially clients themselves" (Anisa, page 14, paragraph 130).

Understanding what has happened to FGM survivors, offering the right support, as well as helping to understand the trauma of the severity of this act, are forms of psycho-education that can lead to preventative work in the future just by exploring the trauma aspect and understanding survivors: "When you've got in touch with the trauma of what's happened to you, if you have experienced FGM, then there's no way you'll do it to your own child. I think that's about opening up conversations rather than adding shame and stigma and fear and threat, and I think, it's a really important way forward". (Elizabeth, page 12, paragraph 74). Talking about the FGM procedure was seen as an important part of therapeutic work. On the notion of understanding the act of FGM as a separate issue and the multi-layered context it is committed under: "...Actually the act of talking about what happens in FGM is a significant part of the therapy" (Elizabeth, page 4, paragraph 18).

There is a need for psycho-education for clients, but also for therapists to be psycho-educated within the work, as each FGM case may be different: "You know, how excited we get when we become aware of FGM. Just again take your time, each is by case-by-case and..." (Maria, page 13, paragraph 102).

Catherine further explores psycho-education, not just for the women who seek therapy but for other people who come from the same country: "Right across the continuum that we need to be teaching more counselling skills to key people in the community and in... In health care, you know, nurses, midwives" (Catherine, page 5, paragraph, 33). Describing trauma and psycho-educating women was one way a therapist was able to explain the concept of FGM or what had happened to survivors: "So it's the education and that awareness that brings up a lot at the workshops" (Jennifer, page 6, paragraph 61).

One of the therapists felt that psycho-education is lacking for therapists working in this field, and explores her knowledge of the lack of research for psychological support: "I think what I've been curious about is that disjoint between what... Again, there's very, very poor data on what the psychological consequences are or the sexual sequelae of FGM are." (Elizabeth, clinical psychologist, page 2, paragraph 12).

Anisa also felt there needed to be more research regarding the future of FGM work: "I think it will be helpful, research-wise, to know what... Statistically, works well for them. And where we shouldn't go or fear towards of what we should do more of" (Anisa, page 14, paragraph 132).

All in all, this theme provided insight into the importance of psycho-education, as well as the lack thereof that has caused issues in therapeutic practice.

Differences in theoretical therapeutic modalities

There were three main therapeutic approaches utilised by all seven of the participants: the psychodynamic, humanistic and integrative approaches. All of the psychodynamic practitioners felt that working psychodynamically was very good for FGM survivors.

One therapist, when exploring working psychodynamically, describes feeling that clinicians working with FGM are aware of boundaries and make them very tight, as clients' boundaries are somewhat non-existent because of their history. Along with being patient with clients, as there is a drop-out rate and it may be difficult for clients to open up about their traumas: "I think we need long term therapy, I think we need to think very carefully about boundaries, boundaries need to be very tight. Obviously there is no boundaries when you're involved with it a hack pieces of it off, there is something about this fear of engagement" (Angelina, page 22, paragraph 88).

Catherine found psychodynamic work to be a positive experience in the therapeutic alliance: "My core way of working is psychodynamically and I think, you know, it's a very effective...... Because it's related to their family history, and it affects their closest interpersonal relationships..." Her lived experience of this modality related to working with survivors' issues within their relationships, in particular sexual relationships: "You know, how and why she feels, you know, the way she does now in relation to the experience of FGM. You know, questions that are... I mean, difficulties in sexual relationships are... Are huge". (Catherine, page 7, paragraph 45).

Maria also describes her experience in working psychodynamically, referring to the importance of the presenting past in therapy: "The theoretical underpinning of that saying is very good for working with FGM women because to just... And the thing of reaching into the past from the person that you are, the piece of work that we did was her having a conversation with that little girl, that little girl who had that African name and had had the FGM." (Maria, page 11, paragraph 121).

Angelina explores her experiences of working with FGM psychodynamically, further discussing the fact that most cases of FGM happen at a young age and that the past does affect the clients' present. Accordingly, she works with issues such as FGM survivors' sense of self-development, trust and interpersonal issues: "Fits perfectly in my eyes… looking at this traumatic experience that you have had in the past and how it impacted on your ability to form relationships and how it's impacted on your ability to trust. How it is impacted on your feelings about yourself". (Angelina, page 12, paragraph 52).

She further attributes her feelings around how it is seen as a 'woman's thing' and the child is not healed: "there is this real kind of suppression and the child is not really healed, emotionally or physically quite often, the child is not healed". Her work is centred around giving the child 'love, care, protection'; a therapeutic holding is described by her. She further explores issues of clients and their primary caregiver dynamics. When therapy is used successfully with clients, she receives feedback that they do feel a lot better: "I often say to clients what would you have said to that child, what would you have liked to have done with that child, and sometimes they come back with rescuing stories, rescue her, or I would like to shout at my mother, or I would like to cuddle her no one cuddled me. And in a way were not physically cuddling them but emotionally were holding them, and giving them the hug that they needed" (Angelina, page 25, paragraph 100).

For therapists working within the humanistic modality, this was seen as the most appropriate approach. A therapist who has a similar cultural background states that a client-centred way of working was most appropriate for her in her experience: "It's very client-centred…" She further describes her way of using the humanistic approach as a sense of going along with the clients' journey, attributed to her own tolerance regarding FGM survivors seeking help for one session and a consciousness of what is comfortable for them: "But what I'm saying in terms of my feelings is that I wouldn't give up on them. If one session wasn't enough I would be happy to come back and..." (Jennifer, page 8, paragraph 95).

Anisa describes the tools implemented from the humanistic approach within therapy, referring to the notion of being in the present moment-to-moment with survivors and their experiences: "So when I was seeing her as for a person-centred, the modality so it suited it very well. So it's literally really being with her in the here and now". She expanded on skills of modelling the client, in particular concentrating on themes of client safety, and projection of exploring the importance of being safe in the therapeutic space: "Modelling someone that will not harm her or has no intention to harm her and wants to be there for her." (Anisa, page 10, paragraph 98). However, while Jennifer implies that using the client-centred approach is impactful, she also notes the importance of understanding that one model does not fit all FGM survivors: "So it's hard to make one kind of model that fits all. It wouldn't make any sense. So for the way that I work in terms of making it client-centred I feel like it gives a lot of validity to the woman" (Jennifer, page 12, paragraph 148). For Sarah, the person-centred approach helped clients with their meaning-making regarding their trauma: "So then it comes back to the whole person-centredness, finding out what FGM has meant to that particular individual and what sense they've made of it and what help" (Sarah, page 4, bottom of paragraph 34).

Therapists having self-awareness was noted as being essential in the work; this was attributed to one of the core skills in the humanistic approach, namely the ability to empathise. Within this work, this led to experiences and feelings of seeking appropriate self-care: "You don't have to go through something to be able to empathise with it… it's about being just very self-aware about you know, what effect that's having on you and using supervision appropriately really". (Sarah, page 10, paragraph 102). Tailoring therapy to what the clients brings to therapy, relating to their experience of the impact of the FGM procedure, and (as mentioned above) the theme of how clients construct their trauma and language they use are all essential to Jennifer when conducting therapy in her modality: "Because sometimes assumptions could be made. Whereas you need to really understand what her views are on it, how does she feel about it" (Jennifer, page 14, paragraph 172). See also Sarah's experience of tailoring therapy within understanding the clients' context of their FGM: "Looking at FGM within the context over the life of [inaudible] might have experienced as well, I think that's so important" (Sarah, page 12, paragraph 11).

One therapist stated that as all cases are different, what may suit one FGM survivor may not suit another: "Yes, and I think also letting go of the idea that we need to see all the patients. I think that's a very old model of therapy".

Furthermore, exploring changes in the current climate of therapeutic work and approaches of working with trauma clients, it could be seen as being quite different in therapeutic work. On the notion of being open to the acknowledgement that not all survivors will want to seek therapy and being able to accept that as practitioners: "You know, so also feeling like supportive work and psychological work can happen in lots of different ways, and not everyone is going to want or need or be able to access, or want, or benefit from what we see as psychotherapy" (Elizabeth, page 9, paragraph 52).

Issues with trust vs. privacy

There were issues that emerged in the work that explicated the lack of FGM clients that pursued psychological support: "I think this idea that there's quite a large population of women who have FGM, but they're not seeking out psychological services" (Elizabeth, page 1, paragraph 8).

One therapist describes the difficulty of providing a safe environment, even though she aims to do that, which impacts her by evoking feelings of frustration as the women find it difficult to trust her: "It is difficult for us to provide a safe environment, but it's quite difficult for us, I think it is part of the frustration that really comes when you talk about feelings that come from this work" (Angelina, page 16, paragraph 66).

In addition to the notion of FGM survivors possibly not trusting mental health practitioners, there were also descriptions of the lack of women coming forward for therapeutic help. A possible avenue for further research is to integrate the hidden population of FGM and work to assist survivors in accessing therapeutic help: "kind of open access and thinking about who doesn't come, and how can we think about interventions for them" (Elizabeth, page 8, paragraph 48,).

Again, the dilemmas involving risk were explored regarding the implications. Issues with clients' trust emerge, as some clients (due to the implications regarding breaking confidentiality if someone is at risk) cause issues in the therapeutic work. On the difficulty some FGM clients may have regarding trusting mental health practitioners: "in terms of confidentiality, we do have to be honest with the women that it will only be taken to a certain extent that if there are any disclosures that's putting anybody at risk, including children...So it's hard to be trusted" (Jennifer, page 3, paragraph, 20).

Sarah also empathises with the theme of trust. Regarding therapeutic modality, there is an element of working in an approach that does not feel too clinical, a feeling of projecting a sense of safety that leads to clients developing trust as the first skill in the therapeutic approach, and the acknowledgement that practitioners guide clients to feeling comfortable: "And so that trust is hugely important...I think much more time needs to be spent engaging them away from kind of clinical settings; it's kind of getting to know them and letting them develop that trust with you first" (Sarah, page 9, paragraph 86).

There is a further acknowledgment among participants regarding the difficulties faced by FGM survivors and their ability to develop trust in the therapeutic relationship. One therapist describes the reality of the work, namely that clients in most instances cannot develop trust: "it's quite impossible, and for her to trust somebody in that little amount of time because you know, she has never been able to trust anybody... I know that you cannot develop trust, realistically how could you develop trust with me" (Angelina, page 24, paragraph 92).

Elements important for future work with FGM clients

Active listening was seen as an important aspect of the work. Maria describes active listening as hearing not only the words but what is actually being communicated, including forms of meta-communication such as body language: "Active listening is really important. You've got to be able to really hear the client and not just hear what's in the narrative, to watch them and see what other signals that you're picking up as well." (Maria, page 9, paragraph 66).

Catherine describes other therapeutic skills she employs in her work with FGM survivors and the importance of her work stemming back to the basics of therapeutic work; active listening was also mentioned in her approach: "basic skills are the most important, which is listening. Listening in an open way. Being non-judgmental. Yes, I mean..." (Catherine, page 9, paragraph 72).

Again, being open and listening to the client's narratives from their world was seen as an important factor. On the ability to name the impact of the trauma and convey congruent and genuine feelings regarding the procedure: "It's a therapeutic act of witnessing and listening, and making okay. That actually by listening to it, with the patient saying it's awful that this thing happened" (Elizabeth page 3, paragraph 16). Elizabeth also discussed being reflective in the work: "That kind of just, just being very, very reflective" (Elizabeth, page 9, paragraph 54). In addition to Elizabeth's acknowledgement of what needs to be taken forward to future work, she also notes the important aspects therapists need to be mindful of regarding future work and explorations focusing on how other therapists can best work with FGM: "just having a holistic perspective, but also modelling that in the way that you run the service,". (Elizabeth, page 9, paragraph 50).

Anisa explores aspects of future work for therapists when working with an FGM client, describing how therapists' ways of being affect the therapeutic alliance and also considering the impact of working with PTSD symptoms. Thus, Anisa discusses assumptions of possible transference in the relationship that may lead to triggering of PTSD: "Is the client going to be intimidated by them? Do they remind them of someone? Because a lot of it is the trauma of what happened and the flashlights and is my therapist going to remind them physically of someone" (Anisa, page 12, paragraph 116).

The experience of FGM client work led to descriptions of the importance of learning and continual professional development, such as being aware of changes in legislation; as the law changed in 1989, then again in 2003, this needs to be studied more: "I think it's important for practitioners to keep up their learning in terms of the changes around FGM. We've got changes in the law in this country" (Jennifer, page 18, paragraph 229).

In addition, for future work, Angelina draws from her experience of being a clinical lead as she describes the importance of women being experienced and having knowledge and passion for FGM work. In her opinion, issues could arise if the practitioner is not an experienced trauma therapist. Her rationale for saying this is that the work is very invasive and the therapist has to be able to withstand it; she has her own doubts about being able to withstand it herself: "You got to have the background knowledge, you got to have a passion with this work, a real deep understanding of how invasive it can be and experienced enough to withstand that. I'm not sure if I am withstanding it, it is having some impact on me" (Angelina, page 21, paragraph 86).

Furthermore, Catherine notes the importance of being open in the counselling room, conveying that another way of not making assumptions is by being open in the work: "I've become more open to the story than perhaps I was at the beginning... You know, sort of, you know, I think about my tone of voice, my use of language" (Catherine, page 2, paragraph 10).

Participants often described the work as a varied experience, as suggested in other themes, what may affect one client may not affect another, so there is consensus that each case is unique: "I feel like every woman that I meet I'm learning something new from. And I think it's important to bring that learning into the work and not assume that," (Jennifer, page 5, paragraph 49).

Catherine describes her feelings about exploring the clients as people rather than the issue of FGM as a whole. Describing her use of elements of empathy within her work: "Yes. So I mean I always think of them as an individual woman. I suppose that's my starting point always when I see someone for a counselling assessment.

Counselling consultations. I think, well here's a woman in front of me and, yes, and I relate to her as a... As a woman really" (Catherine, page 5, paragraph 37).

Again, highlighting an idiosyncratic approach in adding that each case being unique is something important: "I

145

think with the clients that I have worked with, people's experiences are very individual and I think that's the case with any kind of work.... You've got to treat everyone as an individual and you can't come in there with stereotypical assumptions that oh this person's been mutilated so they must be feeling like x, y and z" (Sarah, page 1, paragraph 2).

Importance of self-care

Self-care was deemed to be essential in the work, Maria explores self-care through supervision and personal therapy: "Yes, I really do believe that if you're working with FGM clients, the self-care is really paramount and when I say self-care is you need to be able to talk to someone, so whether you're in supervision or whether you're in personal therapy" (Maria, page 8, paragraph 54).

The importance of practicing self-care by talking to someone who is in the same field, i.e. who is also working with FGM: "Well I've got a colleague that does the same job... I feel like that's a great comfort for me as well because working in this field on your own would be extremely difficult"

(Jennifer, page 8, paragraph 99). For Sarah, as a clinical psychologist, the help she received from her supervisor in relation to the psychological impact of FGM client work led to her experiencing a sense of comfort and being able to explore her impact: "Talking through it with... there's Sudanese clinical psychologists where again it's kind of having that space to kind of just say, gosh that sounded as if it was a nightmare or not necessarily just with her but even kind of amongst my peers as well" (Sarah, page 15, paragraph 160).

In addition to this theme of self-care, there are aspects of the work related to discovering your own limitations as a practitioner: "Like I said, about the women, I've learnt a lot about myself in being able to know how much I can cope with", (Jennifer, page 9, paragraph 111).

Need for supervision/Personal therapy

The therapists also explored their experiences regarding the need for appropriate supervision. In one therapist's account of taking matters she experienced during her work to clinical supervision, she describes the feeling of possibly hiding the fact that her experience of FGM work was impacting her sex life. Issues in being empathetic to FGM clients was also something that was brought up in discussion: "I took it to supervision... It is quite interesting, almost a part of me, even though I am quite open a part of me did not want to talk about sex life in supervision of how appropriate... another therapist was also saying that it was quite similar... was feeling quite guilty that she hadn't been cut, and who were we in the group who were co-facilitators, empathising with these women" (Angelina, page 6, paragraph 28).

Due to the impact of FGM work, one therapist describes other forms of self-care based on the intrusiveness of the work, because of how it impacted her personal life (See theme focused on intrusiveness of FGM work): "One I think is as I said it is very hard to kind of shake off, I think it is something that I could get quite obsessive about. I try not to immerse myself into it, like going off and watching films, or read literature" (Angelina, page 5, paragraph 20). similar". (Jennifer, page 16, paragraph 196).

One therapist had a feeling of despair attributed to the survivor's mothers, noting that she understood the social dilemma while on the other hand having anger towards them to allowing this act to happen to FGM survivors: "It is a kind of mixture of anger and despair really... How could you allow that to happen to your daughters? However, I always see the dilemma in a social context" (Angelina, page 3. paragraph 8).

Chapter X

Reflections & Learnings When Working with FGM

The aim of the present book was to explore therapists' experiences of working with female genital mutilation. Data was collected through semi-structured interviews and analysed through IPA in order to address the following research question: 'How have therapists experienced working with clients who have suffered FGM in childhood, and how might such experience inform future guidelines and practices with this client group?'

Reflections from findings

Most of the therapist interrelated with themes and often had similar positioning. The differences amongst their view was that for some their somewhat understood the as to 'why' this act was done. In particular, there was an understating focusing on the culture of the act and to which context this was done. All reported the difficulty in working with something that does not have much knowledge on, and all focused their opinions around future work. There was a feeling of practitioners as well as clients to be have psycho-education and need to be able to listen to the client stories. Around half of the participant spoke of how it impacted them psychologically such as issues burnout, feeling exhausted and secondary trauma/ vicarious trauma. Due to the variety of culture issues around therapist own identity/culture was another sub-theme that emerged in the work. A myriad of feelings was spoken of and new to my experience there was a feeling of loss for some of therapist. Frustrations in the work were mentioned in particular to women and their drop-out rates. Feelings of guilt emerged in the results because of the resonation all therapist had as been women. Issues of modalities were explored and all therapist believed their own therapeutic frame-work was important in the work. Implications of future work were attributed in therapist experiences and the importance of self-care became a relevant theme. In sum, the results provide an insight to therapist experiences of working with FGM, reflecting on implications, impacts, difficulties and issues in the work.

> **Key Research Findings of therapist experiences of working with FGM survivors 2017**
>
> The significance of understanding the consequences of working with FGM clients, particularly the application to counselling psychology regarding trauma work.
>
> The findings demonstrated potential contributions regarding ways that counselling psychologists/therapists/counsellors may work with this client group.
>
> Understanding the cultural relativist view was essential to the work; having some experience with trauma work was also deemed very important for future therapists to reflect on.
>
> The implications of working with FGM clients led to the inference that self-care, in terms of the right clinical supervision, was essential in therapists' meaning-making.
>
> Overall, this thesis provides clinicians with some insight into possible ways of working therapeutically with FGM survivors.

Key finding 1: The psychological impact of FGM work

There were many factors relating to the psychological impact therapist experienced when working with this client group. Our results reflected constructivist self-development theory and recent research of VT as an experience. Research indicates for trauma work therapist/counselors personal functioning is increased awareness of the reality and occurrence of traumatic events (Trippany, Kress, & Wilcoxon, 2004). For example, this was reflected by Angelia's experience of FGM and often worried about women from that culture being subjected to this act. This reality makes counselors more aware of their vulnerability. As reflected further in Chapter 9 on implications for future work in sub headings of elements of future work. Pearlman & Saakvitne, (1995b) state safety and security are threatened when counselors become cognizant of the frequency of trauma, often resulting in a loss of feeling in control because of hearing clients' stories in which the control was taken from them. In addition, the helplessness of a counselor to change past trauma can challenge, or even shatter, the counselor's identity. This book was reflective of the psychological impact of working with FGM. Just under half of the participants reported feelings of Vicarious traumatization, burnout, parallels with client's helplessness and experiences relating to therapist sense of self and loss of self. Even though participants reflected on the difficulty of working with trauma, FGM was deemed by all a form of trauma based work as reflected in chapter six. Bell (2003) studied therapists who work in the field of violence against women and found that 40% became more grateful for their lives, appreciate their relationships more and are less judgmental. Satkunanayagam, Tunariu and Tribe (2010) spoke about growth through adversity and the "rewards" of trauma work, their participants describing a sense of hope and goodness in humanity. Brockhouse, Msetfi, Cohen and Joseph (2011) conducted a study that aimed to examine the psychological growth of therapists, following the vicarious exposure to trauma. 118 therapists have completed measures of vicarious exposure to trauma, growth, empathy, sense of coherence and perceived organisational support. The results revealed that empathy is a positive predictor for growth. Empathy also moderated the exposure to a growth relationship when growth involved relating to others. These are vital findings in the field of trauma and for those working with clients who have experienced trauma, offering new perspectives on recruitment, training and supervision of the therapists working in this field. This was reflected of the work with FGM clients as sub theme four reflects.

Key finding 2: The emotional impact of working with FGM

There is a dearth of knowledge regarding the emotional impact of working with FGM clients. As this study was an IPA study, a key experience of the participant's subjective experience related to the myriad of feelings such as guilt, sadness, anger. Even though there is no research specifically directed to FGM work, this research focuses on three main feelings guilt, sadness and anger. According to the research that pertains to the psychological impact guilt can be represented through issues such as intimacy with partners may become difficult as guilt and intrusive thoughts related to a client's abuse become present when engaging into intimacy (Saakvtine & Pearlman, 1996). This was shown by one of the participant's when she having sex with her partner and the act of FGM is extremely relevant to issues with intimacy. This further led to some participants feeling guilty and experiencing overwhelming grief in their descriptions of sad feelings, this is reflective of the research done by Herman (1992).

Key finding 3: the cultural dynamics involved in the work

This was most profound in the work, as each experience the therapist explored related to the cultural embeddedness of the act (Pendersen, 1991). As we are aware by the literature this is a culturally embedded act from reasons pertaining to chastity, being accepted by the community and for male sexual pleasure. Even though this reached reflected this, conversely, they were child abuse vs context and issues with risk. As stated there is 14 years of imprisonment if caught or suggestions made if this was bought to the attention of therapist (Lockhart, 2004).

Experiences differed among practitioners as firstly most

of the clinical psychologist and some of the humanistic counsellors did not see FGM as child abuse as they were able to understand the context of why FGM happens, there was an acceptance of this is harmful practice but refrained from seeing FGM under the umbrella of child abuse as in therapist experiences this act is done out of love. On the other hand, some of the participants understood FGM as child abuse and for Catherine she further explores issues of risk of clients seeking asylum and the impact it has on the therapeutic process and relationship. Only one of the many therapist felt they may be a risk, but with all of participant's issues of trust emerged and at times the legislative aspect of FGM work could impact the work negatively. It could be perceived as FGM work as cultural psychology work as you cannot separate FGM work from the cultural dynamics of the practice (Heine, 2007).

Key Finding 4: Therapeutic implications regarding the work

Reflecting through the fourth major theme it was made apparent from all the participants the need for psychoeducation not only for FGM clients seen in therapy put for practitioners to be more culturally aware of the trauma aspect of the work. As Pearlman & Saakvitine, (1995b) of the importance of education and training focused on "traumatology" is vital for trauma counselors and can decrease the impact of VT. Follette, Polusny, and Milbeck (1994), stated that 96% of mental health professionals reported that education regarding sexual abuse was imperative to effective coping with difficult client cases. Chrestman (1995) also found empirical evidence that supported use of additional training to decrease the symptomatology of posttraumatic stress disorder in counselors working with trauma clients. Furthermore, Alpert and Paulson (1990) suggested that graduate programs for mental health professionals need to incorporate training regarding the impact of clients' childhood trauma and its effects on VT. This relates to chapter 6.

From the experiences of self-care aspect of FGM work it was important peer supervision more than half the participants. Peer supervision groups serve as important resources for trauma counselors/ therapist (Catherall, 1995).

Sharing experiences of VT with other trauma counselors offers social support and normalization of VT experiences. This normalization lessens the impact of VT, which in turn amends cognitive distortions and helps counselors maintain objectivity. Other benefits include reconnecting with others and sharing potential coping resources (Catherall, 1995). Pearlman and Mac Ian (1993) found that 85% of trauma counselors reported discussion with colleagues as their most common method of dealing with VT. Peer supervision methods are helpful in providing trauma counselors with validation and support, in providing them with the opportunity to share new information related to therapeutic work, and in allowing them to vent their feelings (Oliveri & Waterman, 1993). Talking to colleagues about their experience in responding to trauma offers trauma workers support in dealing with aftereffects (Dyregrov & Mitchell, 1996). Peer supervision has also been found to decrease feelings of isolation and increase counselor objectivity, empathy, and compassion (Lyon, 1993).

Another form of self-care was clinical supervision. It was

deemed supervision helps alleviate issues of countertransference and traumatic reactions (Rosenbloom et al., 1995). "It is important for caregivers to have a variety of peer support resources to allow easy access to share with others the burden of bearing witness to traumatic events" (Yassen, 1995, p. 194).

Other sub themes focused on issues with trust and privacy in the work and in most research regarding violence against women this is common issue. Regarding therapeutic modality, no specific therapy was identified as the best or most appropriate form of therapy. The results revealed therapist stating the importance of uniqueness of FGM client work and the importance for future therapeutic work or assessment is being aware that idiosyncratic approach is deemed more acceptable when working with FGM clients, for one client a humanistic approach may be suited and for another psychodynamic long-term work, this also led to practitioners stating a need for more research in this field.

Chapter XI

Implications for Clinical Practice

The aim of this book was to explore the experience of therapists working with clients who have suffered FGM in childhood; with the view to both informing future practice with such clients, and offering information to those without experience of working with FGM. An important aspect to reflect on is that even though the title reflected on FGM clients who had experienced this procedure in childhood, however, as therapist accounts denote that in their practice they encountered women from many ages such as having FGM performed for punishment, although majority of FGM cases tend to occur in childhood (Momoh, 2001). The clinical implications to this study led to contribution to knowledge with the counselling psychology discipline.

Implications for clinical practice for the discipline of counselling psychology

Empathise ways of facilitating understanding between therapist and FGM survivors:
- Consideration of the language constructs clients use in meaning making of their trauma.
- Understanding the cultural context of the practice
- Awareness of terminology use: avoiding language that isolates survivors such as references of the procedure being known as child abuse.
- The use of legislation in a harmonious approach, being sensitive and thoughtful regarding issues of safeguarding.

Managing emotions and developing increased self-awareness as therapist:
- Understating the parallel processes within the work
- Developing appropriate strategies of unwanted thoughts and emotions
  (VT, STS, burnout, countertransference, disengagement, frustrations).
- Gaining appropriate self-care strategies

Therapeutic effectiveness within FGM client work:
- By developing increased personal awareness to enhance therapeutic relationship.
- Engaging empathetically with FGM survivors, while ensuring that appropriate boundaries have been maintained and emotional observing is ongoing.
- Continue professional development for training about the FGM procedure and training in trauma work.

This book showed a greater understanding of the experience of therapists working with clients who have suffered FGM in childhood, such as psychological impact in the work, the clinical implications and the cultural dynamics that occur in the work. Therapist made suggestions as to how future practices and guidelines can be adapted to better reflect such experience. For example, this was done by therapist reflecting on experiences of vicarious trauma, issues to mindful of such as being more psycho-educated, being more open and providing clients a safe place by listening to the client's narratives and further being aware that each case is different. So, if one therapeutic modality worked for one FGM survivor that may not be the case for another, thus substantiating an idiosyncratic approach when working with FGM survivors.

This book also provided a clearer understanding of problematic elements of working with FGM in the therapeutic relationship, such as issues with trust and privacy, the legislative element of the work such as naming it as child abuse and issues with possible risk when working with FGM clients.

Therapist also demonstrated an appreciation of some of the effects on therapist experience of cultural and ethical dilemmas and differences, such as understanding the context of the act and therefore being able to understand the client's narratives. However, therapist also mentioned issues with language and difficulty in understanding the many different cultures and at times this could lead to a hindrance. Therefore, therapist advocated in their narratives the importance of seeing individuals from similar cultures giving counselling.

This research could be deemed as informing the development of therapeutic formulations and conceptualisations of working with violence against women, FGM survivors. The key finding in conceptualization is understanding the context of this act and providing a safe-place where the therapist actively listens by providing therapy or assessment in an idiosyncratic manner. In addition, this book could provide insight to the application of relational theory to better represent work with this client group for example emphatic encounters that led to therapist vicarious trauma and becoming activist after their work. Their experiences led to further issues of lack of women coming forward which again could be issues relating to future work in the hidden population research.

In conclusion, this book provided an insight into therapists' perspectives, offering a deeper understanding of the impacts that may be caused by working with trauma and the importance of self-care.

Chapter XII

Closing Remarks & Reflections

Future research

While this book yielded, some common accounts relating to psychological therapists' experiences of working with female genital mutilation, it also served to highlight some areas that may warrant further research. As discussed previously, despite the purposive recruitment strategy of psychological therapists, all of which had worked with FGM and women, it could be claimed that the participant view may be biased as further research might aim to replicate the study with a more focused recruitment strategy targeting other specific types of psychological therapists (e.g. male therapist or the focuses being on a specific form of therapy i.e. only psychodynamic therapist, CBT therapist or just counsellors), to examine any similarities and differences in the processes and experiences identified in the current study.

Due to the many themes that emerged in the work, this research provides other researchers to develop research questions and further understand issues on a deeper level. For example, taking the theme of psychological impact many sub themes emerged in the work, one being therapist experience of vicarious traumatisation this led to another major theme of clinical implications which reflected the importance of self-care. Patterns in the research emerged substantiating evidence of trauma work research (Pearlman, 1995). Therefore, researchers could possibly do a longitudinal study of therapist accounts of working with FGM 3 months after seeing a client, to 6 months, up to 12 months and possibly investigate further elements of self-care as FGM client based therapist.

This study linked to several existing areas of psychotherapy, in particularly reflecting from chapter 9, reflecting the essential ability of 'active listening', something in most therapeutic modalities is deemed as an imperative skill as seen in all counselling work (Cooper & Mearns, 2005; Lemma 2003; Owen, 2009). Further suggestions were made on being reflective and holistic in one's approach when working with FGM clients in the future. Another possible experience of the therapist for further research is this notion of therapist on continual professional development and continuous learning in this field, this may be a potential study using a quantitative measure and psychometrics that could analyse what teaching and courses are out there and how many psychological therapist have sought continual training and if this training was helpful. Furthermore, Catherine in her narratives denotes this need for others from the community to be trained or become mental health advocates for FGM survivors.

Therefore, research in investigating if there are counsellors from the same communities of FGM clients and how they have experienced the work could add to further understanding of how psychological therapist can work with FGM within a therapeutic setting.

Supervision was mentioned by participants to be an important part of their meaning making process in relation to a form of self-care. This varied for participants for some it was vital and for other there were negative experiences (E.g. Anisa). This finding could benefit from more detailed examination, considering the minimal emphasis on the function of supervision in discussing tin the many ways it helped or did not help the work with FGM and in turn the lack of empirical research. Further research examining supervisors' and supervisees' experiences of discussing may provide insight into issues relating to providing better care or help in finding appropriate interventions for FGM survivors.

In sum, the last issue mentioned by participants was the need of empirical research regarding the best psychological support for women. As this research has revealed this can be seen as a form of trauma work based on therapist own experiences of VT, fatigue and need for self-care. Therefore, evidence-based research such a control intervention study such as Eye movement desentization therapy (Brown & Shapiro, 2006) for FGM survivors could possibly be another study to inform the practice of counselling psychology and the most appropriate form of therapy. This study had no differences in therapeutic modality but exposed the importance of using an idiosyncratic approach in assessment and therapy. However, by not having empirical research on the best form of therapy may be considered a hindrance to the work as there needs to further research in analysing the most appropriate form of research based on statistical research.

Reflections on analysing the data: Analyzing participants' transcripts made me aware of the potential to be drawn into some unhelpful processes and provided some insight into how I should approach the data. My initial understanding of what I saw FGM may have impacted the work. Being an activist and a feminist this would lead to the possibility of impacting the data based on my own assumptions and perceptions. The fear was not to have the data as one sided and to be neutral in the work. This was reflected in research supervision, by possible dynamics regarding my own experience of working with FGM and my activist opinion of seeing FGM as a child abuse and violation of women rights, as a factor unconsciously impacting my interpretations. There was an emphasis on bracketing my own views and opinions about participants' content in order to be a more curious and empathic researcher and better engage with the transcripts. It was also important to acknowledge how my theoretical interests could influence how I engaged with the data.

Reflections

A significant motive for my exploration of therapist working with FGM clients was my own personal experience of working with FGM clients in my clinical training. My opening reflections touched on having an activist approach to any violence against women based on a mother who had survived domestic violence. Therefore, research into violence against women was something that I was always going to investigate. However, in my clinical experience of working with FGM survivors, I felt a lack of support as not many supervisors had ever experienced working with FGM survivors and felt overwhelmed with the work. The clinical presentations that I had encountered related to issues of miscarriage, physical complications such as: cysts that grew to the size of a tennis ball, or periods being stuck in the uterus and not being able to release adequate blood flow.

As this book focused on experiences of clinicians, the prominent themes related to patient's mental health descriptions were disclosed. For example, some of the experiences focused on romantic relationships FGM survivors developed. One client described a male sexual partner saying to her she did not please him sexually as she did not have a fully formed vagina. This led to the patient describing her ex-partner as taking pictures from the internet to show her what she described as saying this is 'what a real vagina looks like'. This left the patient mentally scared and led to low self-esteem and issues with her confidence.

Another example by a clinician was of a Caucasian British born lady who was taken to her husband's home country and the act taken out on her, her experiences were described as helpless and hopeless, which is often a parallel process described by clinicians that work with FGM survivors evokes within them. Even I felt helpless and hopeless, when stories of their traumas were described and the issues with relationships they had encountered.

FGM is difficult for all people involved because some women as described in earlier sections of the book see it as circumcision and if their daughters do not have this act done it can be seen as a disgrace to their family, society and have impact on their marital future. Nonetheless, an important element as explored earlier is the need for women to be psycho-educated, about this topic. A key issue that emerged was culture dynamics, as one of clinicians interviewed mentioned how ear piercing was the norm in her culture and stated how lucky she was that she was from a different culture that only subjected young girls to only ear piercing. Whereas, other views related to clinicians understating the context of why FGM was practiced. This was a difficult position to hear and understand, but nonetheless that was an clinicians' meaning making/subjective experience when working with FGM clients. The most important aspect of this work is developing movements to help psychological treatment for survivors. We now currently have knowledge of the psychological and emotional impact FGM treatment has on psychological therapist. We have now touched on issues that emerge in counselling, but more statistical analysis of specific modalities is needed to demonstrate what may be the best form of treatment for survivors. Realistically this is a global issue, that may never be eradicated, but developing knowledge to the best form of treatment is continually needed. This is a hidden population issue and we need to be able access appropriate care and avenues for survivors to access without being judged by their community.

This is a sensitive topic and sensitivity is needed when exploring matters related to treatment, issues of legislative framework, as mentioned in the introduction of the book this is and would be an illegal act, having consequences in the

United Kingdom for fourteen years of imprisonment. It is considered as child abuse (NSPCC, 2018). As practitioners, what we are left with is the complexities in building trust with patients, and issues that may emerge if safe-guarding teams need to be alerted of any future FGM practices or risk relating to children. If as mentioned by some of practitioners the context of the act can be understood, themes of duty of care and breaking confidentiality would be deliberated as ethical dilemmas for practitioners. Nevertheless, the act of FGM in the UK itself has clear laws, therefore mandatory reporting would be deemed essential for any clinician to do, if there is any disclosure of future FGM acts to be committed.

Final points

Initially as stated I wanted to understand client's experiences of therapy, as this was not accepted by university ethics, the therapist experiences of working with FGM survivors was considered. What I have learnt through the experiences of working with FGM survivors is the psychological and emotional impact of the work. This is a different form of trauma and if looking at this through the lens of child abuse the perpetrators are commonly the mothers. Therefore, the anger, hurt pain and sadness often experienced in the transference and countertransference relationship is toward the mothers. However, understanding the cultural paradox is essential in the work. Inevitably as with most trauma work, therapist are at risk of vicarious traumatization as exhibited by the participants in this study. In addition to this I as the researcher exhibited a secondary trauma when therapist re-lived their experiences. This led for my own personal self-care mechanism as going for two hour walks and having three-week gap between each participant interview.

Therefore, what I have observed in this work is the importance of self-care and being aware of our own limitations to any therapeutic work not just work trauma treatment. Whether, FGM treatment can be seen as trauma therapy from this current research that would be deemed appropriate. In essence, this book focuses on many clinical implications but also touches on how therapist experiences can help future clinicians to mindful of when working with this client group.

References

Abdulcadir, J., Rodriguez M, I., & Say, L. (2015). Research gaps in the care of women with female genital mutilation: an analysis. BJOG: An international Journal of Obstetrics & Gynaecology. 122, 3, 249-303.

Al-Sabbagh, L.M. (1996) The Right Path to Health: Health Education through Religion. Islamic Ruling on Male and Female Circumcision 1996 Alexandria, Egypt: WHO Regional Office for the Eastern Mediterranean.

Annas C.L., (1996). Irreversible error: the power and prejudice of female genital mutilation J Contemp Health Law Policy 12 325–53.

Balogun, O.O., Hirayama, F., Wariki, W.M.V., Koyanagi, A., and Mori, R. (2013). Interventions for improving outcomes for pregnant women who have experienced genital cutting. Cochrane Database of Systematic Reviews 2013, Issue 2. Art. No.: CD009872. DOI: 10.1002/14651858.CD009872.pub2.

Behrendt, A., and Moritz, S. (2005) Posttraumatic Stress Disorder and Memory Problems After Female Genital Mutilation, The American Journal of Psychiatry, 162(5), pp.1000-1002.

Bell, H. (2003). Strengths and secondary trauma in family violence work. Social work, 48, 4, 513-522.

Bell, K. (2005). Genital cutting and Western discourses on sexuality. Medical Anthropology Quarterly 19(2): 125-148.

References

Berg, R. and Denision, E. (2012) Does female genital mutilation/ cutting (FGM/C) affect women's sexual functioning? A systematic review of the sexual consequences of FGM/C, Sexuality Research and Social Policy, 9(1), pp.41-56.

Brockhouse, R., Msetfi, M.R., Cohen, K., & Joseph, S. (2011). Vicarious exposure to trauma and growth in therapists: The moderating effects of sense of coherence, organizational support and empathy. Journal of traumatic stress. 24, 6, 735-742.

Brown, J. S., Collins, A., & Duguid, P. (1989). Situated cognition and the culture of learning. Educational Researcher, 18 (1), 32-42.

British Medical Association (2011) 'Female Genital Mutilation: Caring for patients and safeguarding children', Guidance from the British Medical Association. Available at http://bma.org.uk/-/media/files/pdfs/practical%20advice%20at%20work/ethics/femalegenitalmutilation.pdf?la=en [Accessed 8 December 2015].

Burrage, H. (2015) Eradicating female genital mutilation: a UK perspective. Ashgate Publishing Limited.

Cheung, F.M. van de Vijver, F.R., & Leong, F.T.L. (2011). Towards a new approach to the study of personality in culture. American Psychologist. 66, 593-603.

Chung, S. (2015) Telephone interview with Dr Brenda Kelly, 4 December.

Chung, S. (2016). 28 Too Many - FGM lets end it - The Psychological Effects of Female Genital Mutilation: Research blog by Serene Chung http://28toomany.org/blog/2016/may/16/psychological-effects . . .

Cieurzo, C. & Keitel, M.A. (1999) Ethics in qualitative research. In M. Kopala & L.A. Suzuki (Eds). Using qualitative methods in psychology. London: Sage.

Craft, N. (1997). Life span: conception to adolescence BMJ, 315, 1227–30.

Elliot, R, Fischer, C.T., & Rennie, D.L. (1999) Evolving guidelines for publication of qualitative research studies in psychology and related fields. British Journal of clinical psychology, 38, 215-229.

El-Defrawi, M., Lotfy, G., Dandash, K., Refaat, A. and Eyada, M. (2001) Female Genital Mutilation and its Psychosexual Impact, Journal of Sex & Marital Therapy, 27(5), pp.465-473.

Elnashar, A. and Abdelhady, R. (2007) The impact of female genital cutting on health of newly married women, International Journal of Gynecology & Obstetrics, 97(3), pp.238-244.

Epstein, D., Graham, P., and Rimsza, M. (2001) Medical Complications of Female Genital Mutilation, Journal of American College Health, 49(6), pp.275-280.

Equality Now and City University London (2014) 'Female Genital Mutilation in England and Wales: Updated statistical estimates of the numbers of affected women living in England and Wales and girls at risk', Interim report on provisional estimates. Available at http://www.equalitynow.org/sites/default/files/FGM%20EN%20City%20Estimates.pdf [Accessed 8 December 2015].

Farber BA. Introduction: A critical perspective on burnout. In: Farber BA. editor.

Stress and burnout: The human service professions. Pergamon, New York. 1983:1-20.

Figley, C. R. (Ed.). (1995). Compassion fatigue: Coping with secondary traumatic stress disorder in those who treat the traumatized. New York Brunner/Mazel.

Figley, C. R. (1995). Compassion fatigue as secondary stress disorder: An overview. Compassion fatigue: coping with secondary traumatic stress disorder in those who treat the traumatized (1-20). New York: Brunner/Mazel

Festinger, L. (1957) A Theory of Cognitive Dissonance. California: Stanford University Press.

FORWARD (2002) Female Genital Mutilation: Information Pack. Available at http://www.equation.org.uk/wp-content/uploads/2012/12/Forward-Female-Genital-Mutilation-Information-Pack.pdf [Accessed 8 December 2015].

Gee, P. (2011) 'Approach and Sensibility': A personal reflection on analysis and writing using Interpretative Phenomenological Analysis, Qualitative Methods in Psychology Bulletin, 11.

Giorgio, A. & Giorgio, B. (2008). Phenomenology. In JA Smith (ED). Qualitative Psychology: A practical guide to methods (2nd ed). London: Sage.

Halpern, D. (2004). Social Capital. Cambridge: Polity Press.

Harris, P. (2004). The paradox that divides Black America. Observer, 14th December, pp21-2.

Hatfield, E., Cacioppo, J. T., & Rapson, R. L. (1994). *Emotional contagion.* Cambridge: Cambridge University Press.

Heine, S.J. (2007). *Cultural Psychology.* New York: W.W. Norton.

Huggard, P. (2003). Secondary Traumatic Stress: Doctors at risk. *New Ethical Journal.* 6:9-14.

Hussein, L. (2015) We Cannot End FGM Without Supporting Survivors. Available at https://dfid.blog.gov.uk/2015/03/04/we-cannot-end-fgm-without-supporting-survivors [Accessed 22 January 2016].

Inglehart, R., & Welzel, C. (2005). *Modernization, cultural change and democracy: the human development sequence.* New York: Cambridge University Press.

Jones, A., (2012). Working psychologically with female genital mutilation: An exploration of the views of circumcised women In relation to better psychological practice. (unpublished Thesis roar.uel.ac.uk/1437/1/2012_DclinPsych).

Keel, A. (2014) Re: Female Genital Mutilation (Letter to Health Professionals in Scotland). Available at http://www.sehd.scot.nhs.uk/cmo/CMO (2014)19.pdf [Accessed 8 December 2015].

Knipscheer, J., Vloeberghs, E., van der Kwaak, A., van den Muijsenbergh, M. (2015) Mental health problems associated with female genital mutilation, *BJPysch Bulletin,* 39(6), pp.273-277.

Larkin, M., Watts, S., Clifton, E. (2006). Giving voice and making sense in Interpretative Phenomenological Analysis. *Qualitative Research in Psychology,* 3, 102-120.

Lockhart, H. (2004). Female genital mutilation: treating the tears. London: Middlesex university press.

Mahoney, M. J. (2003). Constructive psychotherapy: A practical guide. New York: Guilford.

Maslach, C. (1982). Burnout: The cost of caring. Englewood Cliffs, NJ: Prentice Hill.

Mason, M.A. (2001). The equality trap. Piscataway, NJ: Transaction.

McCafrey M, Jankowska A, Gordon H. (1995). Management of female genital mutilation : the Northwick Park Hospital experience. BJOG: an international journal of obstetrics and gynaecology 102(10):787–90.

McCann, L., & Pearlman, L.A. (1990). Vicarious traumatization: A framework for understanding the psychological effects of working with victim. Journal of Traumatic Stress, 3, 131-149. women in Sweden. Midwifery, 24, 214-225.

McCleary P.H.(2004): Female genital mutilation and childbirth: a case report, Birth, 21 (4):221-223.

Meier ST. The construct validity of burnout. Journal of Occupational Psychology
1984;57:211-219.

Memon, A., (2014) 'Female Genital Cutting: A community based approach to behaviour change', Working Paper.

Momoh C, (2005) ed. Female genital mutilation. Oxford: Radcliffe Medical Press.

Momoh, C. (2010) Female Genital Mutilation. Trends in Urology, gynecology & sexual health. 15, 3, 11-14.

Momoh, C., & Umoren, I. (2016). Female genital mutilation. In J. Payne-James, & R. W. Byard (Eds.), Encyclopedia of forensic and legal medicine (second edition) (pp. 507-514). Oxford: Elsevier. doi:http://doi.org/10.1016/B978-0-12-800034-2.00178-6

Morison L, Dirir A, Elmi S, Warsame J, Dirir S. (2004) How experiences and attitudes relating to female circumcision vary according to age on arrival in Britain: A study among young Somalis in London. Ethn Health. 4;9(1):75–100.

Mulongo, P., Hollins Martin, C., & McAndrew, S. (2014). The psychological impact of female genital Mutilation/Cutting (FGM/C) on girls/women's mental health: A narrative literature review. Journal of Reproductive and Infant Psychology, 32(5), 469-485. doi:10.1080/02646838.2014.949641

Oljira, T., Assefa, N., & Dessie, Y. (2016). Female genital mutilation among mothers and daughters in harar, eastern ethiopia. International Journal of Gynecology & Obstetrics, 135(3), 304-309. doi:http://doi.org/10.1016/j.ijgo.2016.06.017

Pearlman, L. A., & MacIan, P. S. (1995). Vicarious traumatization: An empirical study of the effects of trauma work on trauma therapists. Professional Psychology: Research and Practice, 23, 353-361.

Pearlman, L. A., & Saakvitne, K. (1995). Trauma and the therapist: Countertransference and vicarious traumatization in psychotherapy with incest survivors. New York: W. W. Norton.

Penderson, P.B. (1991). Counselling international students. Counselling Psychologist, 19, 10-58.

Pereda, N., Arch, M., & Pérez-González, A. (2012). A case study perspective on psychological outcomes after female genital mutilation. Journal of Obstetrics and Gynaecology, 32(6), 560-565. doi:10.3109 /01443615.2012.689893

Pilkington, A. (2002). Racial disadvantage and ethnic diversity in Britain. London: Palgrave.

Rashid, M., & Rashid, H.M., (2007). Obstetric management of women with female genital mutilation. The Obstetrician & Gynecologist, 9, 95-101.

Read, J. (2004). Poverty, ethnicity and gender. In J. Read, R. Bentall, & I. Mosher (Eds), Models of madness: Psychological, social and biological approaches to schizophrenia (pp. 161-94). London: Bruner Routledge.

Sanderson, C. (2013) Counselling Skills for Working with Trauma (Essential Skills for Counselling). London: Jessica Kingsley Publishers.

Shell-Duncan, B. (2001). The medicalization of female "circumcision": Harm reduction or promotion of a dangerous practice? Social Science & Medicine, 52, 1013–1028.

Smith, JA. (2004) Reflecting on the development of interpretative phenomenological analysis and its contribution to qualitative research in psychology. Qualitative Research in Psychology, 1, 39-54.

Smith, J.A. (2011) Evaluating the contribution of interpretative phenomenological analysis. *Health Psychology Review*, 5, 9-27.

Smith, A.J.M., Kleijn, W.C.H.R., & Hutschemaeers, G.J.M. (2007). Therapist reactions in self-experienced difficult situations: An exploration. *Counselling and Psychotherapy Research*, 7, 34-41.

Smith, J, A., Flowers, P., and Larkin, M. (2009) *Interpretative Phenomenological Analysis.* London: Sage.

Smith, J.A and Osborn, M (2008) Interpretative phenomenological analysis. In J.A Smith (ed) *Qualitative Psychology: A Practical Guide to Methods.* London: Sage. (2nd ed).

Stamm, B. H. (1997). Work-related secondary traumatic stress. *PTSD. Research Quarterly*, 8, 2. (http://www.ncptsd.org/publications/rq/rq-lrst.html).

Stiles, W.B., (1993). Quality control in qualitative research. *Clinical Psychology Review* 13: 593 618.

Sureshkumar, P., Zurynski, Y., Moloney, S., Raman, S., Varol, N., & Elliott, E. J. (2016). Female genital mutilation: Survey of paediatricians' knowledge, attitudes and practice. *Child Abuse & Neglect*, 55, 1-9. doi:http://doi.org/10.1016/j.chiabu.2016.03.005

Tobin, W, T., & Jaggar, M.A., (2013). Naturalizing Moral Justification: Rethinking the method of Moral Epistemology. *Metaphilopshy, 44, 4,* 409-439.

Toubia, N. (1994) Female Circumcision as a Public Health Issue, The

New England Journal of Medicine, 331(11), pp.712-716.

Vloeberghs, E., Knipscheer, J., van der Kwaak, A., van den Muijsenbergh, M. (2012) Coping and chronic psychosocial consequences of female genital mutilation in the Netherlands, Ethnicity & Health, 17(6), pp.677-695.

Widmark C, Leval A, Tishelman C, Ahlberg BM. (2010) Obstetric care at the intersection of science and culture: Swedish doctors' perspectives on obstetric care of women who have undergone female genital cutting. Journal of Obstetrics & Gynaecology, 30(6):553–8.

Willie, C.V., Perri Rieker, P., Kramer, B.M., & Brown, B.S. (1995). Mental health, racism and sexism. Pittsburgh, PA: University of Pittsburgh Press.

Whitehorn, J., Ayonrinde, O., & Maingay, S. (2002) Female genital mutilation: Cultural and psychological implications, Sexual and Relationship Therapy, 17(2), pp.161-170. Available at http://www.tandfonline.com/doi/pdf/10.1080/14681990220121275?instName=University+of+Oxford [Accessed 8 December 2015].

World Health Organization (2008) Eliminating female genital mutilation. An interagency statement. Available at http://apps.who.int/iris/bitstream/10665/43839/1/9789241596442_eng.pdf [Accessed 8 December 2015].

Internet Resources

American Psychological Association, (2015). Facts about woman and trauma: http://www.apa.org/about/gr/issues/women/trauma.aspx.

Brown, E., & Hemmings, J. (2013). The FGM initiative: Evaluation of the first phase (2010-2013). Esmeefairbirn foundation. esmeefairbairn.

org.uk/.../The_FGM_Initiative_Final_Report_2013_1.pdf
FGM Act (2003) http://www.legislation.gov.uk/ukpga/2003/31/contents.

Health and social care information centre (2015): http://www.hscic.gov.uk/article/2021/Website-Search?productid=17885&q=fgm&sort=Relevance&size=10&page=1&area=both#top
Plan (2015): http://www.plan-uk.org/because-i-am-a-girl/female-genital-mutilation-fgmProhibition of female circumcision Act 1985: http://www.legislation.gov.uk/ukpga/1985/38/contents

World health organizations (2014) : http://www.who.int/mediacentre/factsheets/fs241/en/.

World health Organizations, (2012). http://www.who.int/reproductive-health/publications/fgm/en/.

UNICEF (2014) Child protection from violence, exploitation and abuse: http://www.unicef.org/protection/57929_58002.html

Lightning Source UK Ltd
Milton Keynes UK
UKRC032232170822
407467UK00003B/180